CULT OF THE
DEAD COW

CULT OF THE DEAD COW

How the Original Hacking Supergroup Might Just Save the World

JOSEPH MENN

PUBLICAFFAIRS

New York

PublicAffairs
Hachette Book Group
1290 Avenue of the Americas, New York, NY 10104
www.publicaffairsbooks.com
@Public_Affairs

Printed in the United States of America

First Edition: June 2019

Published by PublicAffairs, an imprint of Perseus Books, LLC, a subsidiary of Hachette Book Group, Inc. The PublicAffairs name and logo is a trademark of the Hachette Book Group.

The publisher is not responsible for websites (or their content) that are not owned by the publisher.

Print book interior design by Amy Quinn.

Library of Congress Control Number: 2019935886

ISBNs: 978-1-5417-6238-1 (hardcover); 978-1-5417-6237-4 (ebook); 978-1-5417-2442-6 (international)

LSC-C

10 9 8 7 6 5 4 3 2 1

For pulmonologist Dr. Tze-Ming (Benson) Chen,
who saved my life after Def Con 2014

CONTENTS

Photo insert between pages 116 and 117

AUTHOR'S NOTE

ECHNOLOGY IS DECIDING the fate of the world, and we are everywhere in its chains. Electronic surveillance, cyberwarfare, artificial intelligence, and manipulated social media are on the brink of pushing societies beyond a point of no return. Even those of us who saw this coming did not think it would get this dire this fast, and definitely not in this way.

For the past two decades I've covered the tech industry as a journalist, and I have been drawn most often to the issues of security and privacy. They immediately cross lines from business to politics and challenge our ideas about safety, freedom, and justice, and it has been fascinating to watch and occasionally participate as governments, companies, and civic-minded people grapple with the fast-changing ramifications. Security is about power. And it has been getting increasingly complex since the moment the internet escaped from its controlled university environment in the 1980s.

As I worked on my first book out of Silicon Valley, about the rise and fall of Napster, I began to grow more concerned about computer security, or the lack of it. Shawn Fanning was one of the first hackers to be admired by the public at large, and he got early help from a more experienced crew, including some people I kept in touch with and who appear in this volume. Though the record industry would beg to differ, most of Fanning's group were the good guys, tinkering in order to learn, not to be malicious. But all of the trends they pointed me to were bad.

As the state of security deteriorated and the stakes rose, I devoted my next book to the topic. *Fatal System Error* showed the scale of the danger, looking especially at how organized crime and some of the world's most powerful governments were collaborating to leverage inherently flawed technology, the failure of the market for security products, and minimal regulation. At the heart of that book was a true tale of Russian intelligence collaborating with criminal hackers, a scenario that went from shocking at the time of publication in 2010 to widely accepted today.

Since then, many books have tackled the military-internet complex, intelligence gathering, and cyberwarfare, together with WikiLeaks, Edward Snowden, and the 2016 US election. Missing in all of them has been a compelling account of the people dedicated to information security who are out of the spotlight or even in the shadows, fighting to protect our personal data and freedom as well as our national security. In many cases, these people are more colorful than their adversaries. That is especially true of the people whose tale is told in this book: key members of the Cult of the Dead Cow, who have played a role in all of the major issues cited above. While their more overt antics drew attention in the past, until now no one has heard their real story, and some young hackers haven't heard of them at all. Yet the Cult of the Dead Cow is a skeleton key for the whole saga of modern security, especially the struggle to sort through what is ethical. cDc stands in here for many others who are doing heroic work well away from public view.

Fatal System Error was a dire warning during a time when many were oblivious. Now, in a time of wider moral crisis in technology, this book is a rare message of hope and inspiration for tackling worse problems before it's too late.

Joseph Menn

> THE PLAYERS

Cult of the Dead Cow

Kevin Wheeler / Swamp Rat
Bill Brown / Franken Gibe
Psychedelic Warlord
Carrie Campbell / Lady Carolin
Jesse Dryden / Drunkfux
Paul Leonard / Obscure Images
Chris Tucker / Nightstalker
Dan MacMillan / White Knight
Misha Kubecka / Omega
John Lester / Count Zero
Luke Benfey / Deth Vegetable
Sam Anthony / Tweety Fish
Peiter Zatko / Mudge
Laird Brown / Oxblood Ruffin
Josh Buchbinder / Sir Dystic
Christien Rioux / Dildog
Adam O'Donnell / Javaman
Jacob Appelbaum / IOerror
Kemal Akman / Mixter
Patrick Kroupa / Lord Digital

cDc Ninja Strike Force

Chris Wysopal / Weld Pond
Window Snyder / Rosie the Riveter
Limor Fried / Lady Ada

Legion of Doom

Chris Goggans
Scott Chasin

Masters of Deception

Elias Ladopoulos / Acid Phreak
Mark Abene / Phiber Optik

@stake

Alex Stamos
Rob Beck
David Litchfield
Katie Moussouris

CHAPTER 1

> AN EVENING IN SAN FRANCISCO

O N A TUESDAY evening in October 2017, three dozen friends and acquaintances gathered in the San Francisco townhouse of security engineer Adam O'Donnell for a political fundraiser. Though a boom in Bay Area real estate put the hillside place in Glen Park out of the reach of most Americans, it was modest by local standards. There weren't nearly enough chairs for those who came to the dinner party, and the guests made their own tacos and drank wine from plastic cups as they stood. Adam was no swaggering Silicon Valley executive. The Philadelphia native had bought the property before the latest housing boom, using money from the sale of a security company where he had worked to Cisco Systems. Adam had joined the target company when it bought the start-up he had cofounded in 2009, which had been early to take advantage of what became known as the cloud, protecting computers from viruses more quickly than rivals. Adam now moved nervously through his home, thanking guests for coming and redoing the math in his head in hopes that the $250-per-head minimum would make it worth the candidate's plane trip.

Adam wasn't accustomed to entertaining people he didn't know well. Now approaching forty, he'd grown up a working-class kid who liked to tinker and eventually had earned a doctorate in engineering. Even as hacking became the stuff of countless headlines, controversial elections, and undeclared warfare, Adam stayed in the background.

At Cisco, Adam was working on a rare joint effort with Apple to help companies protect employee iPhones. It wasn't particularly glamorous. His most exciting work was something he didn't talk about. Under the handle Javaman, Adam was a longtime member of the oldest, best-known, and most important hacking group of all time, the Cult of the Dead Cow. Walking in Adam's front door, some old-school hackers saw the cow skull hanging in the foyer and got the reference. If not, Adam didn't explain.

[x x]

Though it has never had more than twenty active members at a time, cDc has multiple claims on history. As it evolved from a pre-web community into something like a hacker performance-art troupe, cDc members started the first hacker convention to invite media and law enforcement. They developed hacking tools that are still being used by criminals, spies, and professional network administrators. And they invented the term *hacktivism*, which the group defined as hacking in defense of human rights. It rarely inducted new members, and when it did, cDc usually picked people already established through other groups, making it a supergroup in the rock-and-roll sense—a band formed of people from other bands. As cDc matured, its members became leaders in changing hacking from a hobby to a profession to a mode of warfare, or really several modes. That warfare has metastasized in the past decade, encompassing the US-led Stuxnet attack on Iran's nuclear program, Russia's blackouts of electrical systems in Ukraine, and China's methodical pillaging of Western trade secrets. The unstoppable, semi-automated propaganda that helped propel the 2016 election of Donald Trump was just the latest, most complicated, and most effective twist. Such information operations and sabotage threaten to continue indefinitely around the world with little oversight.

Most Cult of the Dead Cow members have remained anonymous, although sixteen have agreed to be named for the first time in these pages, including all of the previously cloaked core participants. That invisibility, dating to the group's founding in 1984, enhanced its mystique. It also gave the fifty or so sometime participants more freedom to navigate the world without being judged or misjudged, in some cases reaching powerful positions. Yet a few have become not only public but famous over the years, including Peiter Zatko, known online as Mudge. In Boston, Mudge fronted the pro-security or "white hat" hacking group called the L0pht (pronounced "loft"), pioneers for warning software companies about security flaws in their wares, rather than just exploiting them to break into users' machines. Then Mudge's squad turned the L0pht into the first big consulting group of star hackers, called @stake; later he led the cybersecurity efforts at the Defense Advanced Research Projects Agency (DARPA), powering both US military defense and still-undisclosed offensive hacks that headed off worse violence in the Middle East. Even more famous in recent years has been Jacob Appelbaum, alias IOerror. The charismatic American face of Tor, the most important tool for preserving privacy on the net, Jake served as one of the last loyal aides to WikiLeaks leader Julian Assange, and he personally revealed hacking tools developed by the National Security Agency. When his own acolytes exposed Jake for sexual harassment, the Cult of the Dead Cow publicly booted him out. But probably the most influential cDc member in steering hacker culture is Laird Brown, known to most by his handle, Oxblood Ruffin. The father of hacktivism, Laird invented facts and was closer than his followers realized to Western intelligence figures, but he drove moral considerations to the heart of a global debate and ended up saving countless lives.

Because they were the first to grapple with many ethical issues in computer security, cDc members inspired legions of hackers and professionals who came after them. cDc figures and those they trained have advised US presidents, cabinet members, and the chief executives of Microsoft, Apple, and Google. And as issues of tech security became matters of public safety, national security, and ultimately the future of democracy, the Cult of the Dead Cow's influence figured in critical decisions and national dialogue, even if many were unaware of its role. In the Silicon Valley of 2018, cDc shared indirect responsibility for rank-and-file engineers citing human rights to protest their own companies' work with immigration enforcement, the Pentagon, and China.

[x x]

Adam had contributed to other political campaigns, especially in the wake of Trump's election, including some Democratic neophytes identified by the entrepreneur founder of a new Bay Area grassroots group called Tech Solidarity. And he would soon write a program to help target likely Democratic voters on Facebook the way Trump had gone after Republicans. But playing party host was a bit scary for an introvert like him. So Adam had asked one of the Cult of the Dead Cow's most prominent protégés to join him as cohost—Facebook's chief security officer, Alex Stamos. The grandson of Greek Cypriot immigrants who ended up in Sacramento, Stamos had a trajectory similar to Adam's—public schools, serious technical higher education, and then jobs as a principled hacker. One of his first was at @stake, working for Mudge and others in the L0pht who had wowed him by testifying to Congress in 1998, under their hacker handles, about the dismal state of cybersecurity.

Following in cDc's footsteps, Stamos had earned a reputation for independence. When Edward Snowden leaked files showing that the NSA was collaborating closely with the big internet companies, especially to scoop up data on people in other countries, Stamos gave a heartfelt talk on ethics at the biggest hacking conference, Def Con. He declared that despite the lack of widely enforced moral codes, security experts should consider resigning their posts rather than violate human rights. For all the stridency, Yahoo hired Stamos as chief information security officer, part of the general public response by Silicon Valley giants to the exposure of complicity. He stayed until 2015, when he quietly quit over the company's unannounced searches of all user email under a secret court order. Since then he had held the top security job at Facebook, trying to limit the damage of Russian hackers spreading

hacked Democratic emails under false pretenses and fighting other battles against propaganda, despite lukewarm support from above.

Separately from his work at Facebook, Stamos engaged in electoral
politics. At Yahoo, he had briefed Congress on security issues, and he
had been impressed by some representatives and dismayed by others.
Realizing that his seat at a big company gave him special access, he
used that and personal donations to candidates from both parties, including Texas Republican Will Hurd, to push on the issues he cared
about. His legislative wish list included combining US cybersecurity
defense in one agency, instead of having multiple agencies working
mainly on offense. He also wanted to reform hacking prosecutions, currently guided by the sweeping Computer Fraud and Abuse Act, and
prohibit built-in government back doors for spying in tech products,
which Stamos thought would cripple American companies as other
countries turned away. And like former White House cybersecurity advisor Richard Clarke, he wanted a more robust White House process
for deciding what software flaws to hoard for offense and which to disclose for defense. At Facebook, Stamos was quietly helping with special
counsel Robert Mueller's investigation into Russian meddling during
the 2016 election.

Adam figured Stamos would want to support tonight's candidate
because of his technological philosophy and the potential significance
of the race to the future of the country. There were deeper reasons as
well, including a chance to pay a sort of cosmic Silicon Valley pen-
ance. The candidate was Beto O'Rourke, a Democrat who was hoping
to emerge from the primary and face Republican Ted Cruz in November for a Texas seat in the US Senate. Cruz was the heavy favorite
against pretty much anyone. No Democrat had won a statewide Texas
vote since 1994, and Cruz was one of the best-known and best-funded
members of the Senate, the Republican runner-up when Trump won
the national primaries in 2016. But Cruz also had a special resonance
for anyone deeply informed about Facebook, the Mueller probe, or
both, as Stamos was. Cruz once had been the top political client of

Cambridge Analytica, which had siphoned off Facebook data on as many as 87 million mostly unwitting users as it coached Cruz, and then Trump, on how to target them with effective ads. Looking at the full electoral picture, Republicans held a slim Senate majority, and flipping just two seats would allow Democrats to block automatic approval for Trump's Supreme Court and cabinet picks and, if necessary, protect Mueller's probe.

It wasn't just those who had failed to supervise the mindless algorithms at Facebook, Twitter, and YouTube who had something to regret after the 2016 election. The Cult of the Dead Cow had amends to make as well. It had turned the creativity and antiestablishment antics of the hacking world against the mainstream media, hustling national television and print outlets for fun and to raise awareness of various issues. A side group cDc called the Ninja Strike Force, created in innocence but later left unsupervised, had deteriorated and recently attracted race-baiting provocateurs who adopted cDc's methods but not its message. A few latter-day members stirred up hate on social media and promoted the technologist behind the biggest neo-Nazi publications, which actively supported Trump.

After a few words from Adam and Stamos, O'Rourke spoke to the group. He had run a small software company and alternative publication before winning an underdog race for city council and another for Congress, where he was serving his third and final two-year term. Slim and six-foot-four, he wore an open-collared shirt and a blue suit as he explained that he had decided to run on the night Trump was elected president. He and his wife, Amy, had been trying to decide what to tell their three children in the morning, and what they would tell them in later years. "What did we do? How did we account for ourselves?" O'Rourke recalled the conversation. He would have to stand down as a representative to appear on the ballot for the Senate, but O'Rourke had decided it was worth the risk. He had been driving to every county in Texas, his campaign was gaining real momentum, and he thought he had a chance. Education, access to health care, and jobs were more important, he said, than blue or red, and the willingness of voters to install someone who would "blow up the system," like Trump, could be harnessed. The biggest challenge was getting people to the polls.

It helped, O'Rourke said, that Texans hate phonies, so he didn't hide that he opposed Trump's planned border wall, thought Trump should be impeached, and supported abortion rights, the legalization of marijuana, and gun control, as did most Bay Area tech workers. He was already fighting in the House to overrule Trump's Federal Communications Commission and restore net neutrality, which kept internet access providers from favoring some content over others. O'Rourke didn't have to contrast his frankness with Cruz's flexibility. Everyone there knew the incumbent had declined to endorse candidate Trump after he attacked Cruz's wife's looks and suggested Cruz's father had been involved in John F. Kennedy's assassination, before Cruz rolled into line anyway. "We've just owned everything that we are about and believe in," O'Rourke said. Declining money from political action committees hurt, but Adam and Stamos's fundraiser helped. Several who attended it went on to hold their own fundraising parties in a chain reaction. Across the country in Boston, cDc stalwart Sam Anthony, a Harvard doctoral candidate working to make self-driving cars safer, held a fundraiser for O'Rourke that likewise inspired additional East Coast donations.

Though many others would also gravitate toward helping O'Rourke as he gained steam, won the 2018 primary, and drew almost even with Cruz in the polls, the early support in San Francisco and Boston was fitting. Those two cities had the most cDc members. And, as it happened, the group had had its start in O'Rourke's home state of Texas.

CHAPTER 2

> TEXAS T-FILES

IKE MANY OF the internet's earliest adopters, Kevin Wheeler willingly struggled to master the new and clunky medium out of a deep need for human connection. The nerdy son of a university administrator and a music teacher had enjoyed a group of similar friends in Kent, Ohio, where they played Dungeons & Dragons. But then the family moved to Lubbock, Texas, in 1983, and the thirteen-year-old had the culture shock of his life.

It was bad enough just being a rebellious teenager in the heart of the Reagan Republican era. But now, at his new junior high school, Kevin was lost among the culturally conservative evangelicals whose idea of a rebel was hometown hero Buddy Holly. Kevin tried to talk to the rich kids, but they were snobby and mean. He tried the poor kids, and they shocked him, trading tales of sex and drugs. But they let him sit with them, so he stayed.

A couple of other kids had parents working at the big Texas Instruments plant and were also technologically inclined. Others started paying attention to what could happen with computers after seeing the movie *War Games*, which came out the year Kevin arrived in town. The film depicted teenager Matthew Broderick dialing out randomly through a clunky gadget called a modem that sat between his computer and his home phone line. Broderick's character accidentally tapped into a military supercomputer. The budding hackers of Lubbock weren't looking

for trouble either. A couple of the older kids had set up electronic fo-
rums known as bulletin boards, where strangers, using modems to call
in over regular phone lines, could read or leave messages and text files,
which the locals also called t-files. Widespread use of web browsers
was still a dozen years away.

Kevin had put in two years on his Apple II by the time he moved
to Lubbock, so he found the local bulletin boards in short order. There
weren't a lot in his 806 area code, and most were run by hobbyists
talking about computers. Some older teenagers had one that was more
freewheeling, and Kevin and a group of friends chatted there for a
while, until the bigger kids got tired of the hangers-on and banned
them. Kevin was indignant. "We have to make our own and truly be
elite," he told friends. Kevin and the others started several boards and
filled them with text files on heavy metal and parodies of *Star Wars* and
other pop culture topics, as well as satires of the more serious bulletin
board operators and swaggering hackers. The boards cross-referenced
each other's titles and phone numbers and banded together under the
name Pan-Galactic Entropy.

To dial interesting bulletin boards outside the area code meant
hefty long-distance charges on the home phone bill. Anyone without
rich and forgiving parents needed someone else's credit card, or a five-
digit code from a long-distance company like MCI, or some actual
hacking ability. The easiest of those to come by was a five-digit code,
which could be cracked by hand with repeated trial and error by those
who were truly dedicated. The winning digits spread like hot gossip
in the school lunchroom and by bulletin board postings at night. That
worked until too many people used them and MCI noticed and revoked
the number, which would usually take about a month. Then a new one
would be discovered and passed around.

If you spent enough time at it, you could find a bulletin board with
just your kind of content and just your kind of attitude. Most boards let
you download what they had and repost it on your own board, if you had
a modem that was fast enough or that you could let run all night to di-
gest a big file—that is, if nobody needed to make a regular call, so you

could stay connected. Kevin's parents didn't seem to mind his occupying the phone line and staying up late downloading files.

Like many his age, Kevin hunted for new programs he could run on his Apple, which meant obtaining and trading "cracked" versions with the digital controls limiting usage removed, known as warez. But reading and soon writing text files were what Kevin cared most about. It was a creative outlet for him, and he had an audience. He wanted his text files to be funny, or at least provocative, so he could connect to other kids who got the same jokes. After a 1985 summer job at a computer store brought in enough money for Kevin to buy a $715 hard drive, he launched his own bulletin board, Demon Roach Underground. One of the first files to go up was Kevin's nonsense riff on the established genre of subversive files with material like that in the printed tome *The Anarchist's Cookbook*, which gave instructions on most things dangerous and illegal.

Kevin's offering was "Gerbil Feed Bomb," and it used numbered instructions and advised readers to, among other things, grind pet food pellets up, pour the grains into a glass jar, and dump them out again. Then they were to pour gas into the jar, light a fuse, and run away screaming. It was passable juvenile humor. But while it made fun of anarchist credos, it also mocked the police who would respond to the explosion: "The police are your friends!" And it talked about how much fun it was to whack a bag of pet food with a bat and pretend it was Republican first lady Nancy Reagan, inventor of the "Just Say No" anti-drug campaign. Kevin himself was never interested in drugs or even beer, but that didn't mean the Reagans didn't deserve to be mocked.

Online, everyone needed a handle. Kevin picked Swamp Rat because he loved playing in the marsh near his home. The nickname soon evolved to the more distinguished Swamp Ratte and eventually to Grandmaster Ratte. One of his earliest online cohorts took the unoriginal name Sid Vicious after the most pathetic, drug-addicted member of one of the first punk bands, the Sex Pistols. In reality, Sid was an eighth grader named Brandon Brewer who lived in the nearby town of Friendship. Unlike the pale and reclusive Kevin, Brandon played sports. But he and his older brother Ty, known as Graphic Violence, also ran a bulletin board called KGB, after the Soviet spy agency. It hosted real bomb-making instructions, among other things. But it also kept the brothers from getting drunk and getting in trouble outside their house. Kevin later told a friend that KGB "had some nutty retardo sex & violence stuff and some kinda phreaking thing about MCI," referring to the telephonic equivalent of computer hacking.

Brandon had more technological ambition than Kevin. He went dumpster diving outside big company offices, looking for anything that would help him break electronically into those businesses. He also used "blue boxes," which were prime devices for phreaking. They emitted tones over phone lines to rig free long-distance calls. A favorite game was to keep transferring calls to stations farther along in the same direction, eventually circumnavigating the world to ring a second phone in his own house. Such phone tricks were still easier than programming, though the great transition was coming soon. When the Brewer boys got new software for their computer, it still had to be keyed in by hand. One would dictate a line of code while the other one typed. When their fingers grew sore, they switched.

Brandon and Kevin didn't want to seem as menacing as the serious hackers, the ones who might go to jail. "In our circle, there was nothing malicious; you never went in there trying to harm somebody's system," Brandon said. "It was all about getting through the wall." Still, they wanted to be taken seriously. And the name Pan-Galactic Entropy didn't sound menacing enough to be cool. It was too *Hitchhiker's Guide* nerdy. They kicked around possible new names for their effort to tie together their small community of bulletin boards and writing and decided that something with the word *cult* would be sinister and mysterious enough. Cult of what? Some words were too silly, like *strawberry*. But a little silly would be good. "We wanted it to be weird," Brandon said. "Just trying to thumb our nose at the establishment." This was a place with a clubhouse feel, the liberal arts section of the hacker underground. Kevin thought of a creepy hangout nearby, an abandoned slaughterhouse, the unpleasant hind part of the most iconic Texas industry. In that moment, he hit upon the name: the Cult of the Dead Cow.

Though Brandon helped come up with the name and provided hacking prowess, the latter got him in trouble before long. It was the phone phreaking that did it. A friend of the Brewers was house-sitting, spotted an MCI calling card, wrote down the number, and shared it with the brothers, who dialed into other bulletin boards and told readers there to check out the Brewers' board. After too many unexplained charges showed up on the bill of the card's owner, he called police, who visited the friend, who named names. Soon men wearing suits were in the living room at the Brewer house. Thirty-odd years later, Brandon said he wasn't sure exactly what happened next. Maybe the men took the computer away as evidence. Maybe his father threw it in the trash. Either way, that was the end of Sid Vicious. Brandon Brewer went to high school and discovered girls.

Brandon left the group before even meeting the kid that Kevin considers the third founder of the Cult of the Dead Cow, a boy calling himself Franken Gibe who frequented some of the same boards as Kevin. His real name was Bill Brown. In the spring of 1986, Bill called the number for Demon Roach Underground. The board itself didn't appear, because Kevin was working on the software to make it function more smoothly or look stranger. Staring at an empty screen with a prompt, Gibe typed in "hello" and hit Return. "Who are you?" appeared on the next line. Bill hoped he was communicating directly with the system operator, or sysop. He was in luck; it was Kevin. They chatted for a while. Eventually, they worked out that they lived only a few blocks apart and got together in person.

Bill was more into fringe culture—UFOs, secret societies, and B movies—than writing computer code. After *War Games*, he had to beg for a computer from uncomprehending parents who would not even get a telephone answering machine until the late 1990s. "I knew nothing about computers," he said. "What I liked was the idea of a bulletin board, this pre-internet, glorified shortwave radio network." Both boys were outsiders in Lubbock in cultural taste and also within an early internet scene that celebrated hacking feats. They were like the early punk rock bands, who weren't going to be quiet just because they couldn't play their instruments well.

Avoiding the assigned work in his Catholic school, Bill helped mythologize the Cult of the Dead Cow along pseudoreligious lines by drafting an epic "Book of Cow" as his first text file. It was inane and sublime, a 1,100-word running gag on both testaments of the Bible. "The beast rumbled forth, and all was cud and the effluvium of animal. This was the beginning. And from the Moment of the Cow was born all that we call earth," reads a section near the start. Toward the close: "So did the Cult spring in those barren times, and so did fertile minds harvest the crops of justice and truth. The Cult unsheathed the shining blade of knowledge, and into battle marched, resplendent in the dazzling garb of ideals."

Later, Bill reflected on the fact that he had arrived on the hacking scene as a sort of court jester. "I took my stupidity very seriously, and chafed under the oppressive hierarchy of the Informed Aristocracy," he wrote. "Before cDc, there were the Elites and the Losers. It was a simple, feudal, pre-pubescent system of class discrimination, based on connections (primarily) and knowledge or experience in the h/p [hacking/phreaking] arts. . . . cDc was really a liberating force." After a while, Kevin and Bill decided the group couldn't be all ridiculous humor and overwrought exhortations, that it needed some hacker credibility. And so it was that the decidedly untechnical Bill went to the Texas Tech library, studied a book on Unix operating systems, and posted a decent summary of software commands that continued to circulate online for years.

[x x]

Most files back then were computer-language cheat sheets or pieces that taught readers how to connect and where, often for free. But they didn't go anywhere after that. Bill pushed a cDc ethos with "telecom as a means, not an end." The kids' humor punctured any self-importance the group had and made it approachable. cDc slowly absorbed other boards and linked up with still more further afield, including ones run by an El Paso teen with the handle Psychedelic Warlord and someone in Michigan called G. A. Ellsworth, whose real name was Matt Kelly. Both contributed their own text files to the mother ship and were inducted as members. Published from 1987 until 1990, Warlord's eight cDc files included transcribed lyrics by the funny punk band the Dead Milkmen. There was a fantasy about visions driving the narrator to murder: "No longer could this strong desire in my mind be suppressed. Recognize this fact, my one and only goal in life became the termination of everything that was free and loving." The first cDc file Warlord published, the year he turned fifteen, asked readers to imagine a better world, or at least a better country, without money. After a nonviolent end to the government, he foresaw the end of starvation and class distinctions.

Another file Warlord submitted, the following year, was a transcribed interview with a self-proclaimed neo-Nazi who maintained that Hitler was misunderstood and didn't personally want Jews killed. Warlord and a Jewish friend questioned the man about his theories and let him ramble. After the interview, Warlord wrote in the cDc file, "We were trying to see what made him think the horrible things that he did." He added that he was opposed to censorship, so if people wanted to learn more about the man and his Aryan church, they could write to his post office box in El Paso. He hoped readers would inundate it with messages or counterarguments, or just antagonize the guy. "Surely they'd appreciate some 'fan' mail," he wrote.

Though his family lived comfortably and were considered high status, Warlord felt like a misfit. He too abused phone cards and downloaded pirated games. "When Dad bought an Apple IIe and a 300-baud modem and I started to get on boards, it was the Facebook of its day," he said. "You just wanted to be part of a community."

By recruiting leaders of other boards, cDc began to act a little like the supergroup it would become a decade later. But in those simpler days of the late 1980s, the main criteria for membership in cDc were the following: (1) be known to an existing member, (2) don't be boring, and (3) don't be an asshole. A girl who went by "Lady Carolin," actually named Carolin (Carrie) Campbell, got to know Warlord from his board and then joined cDc at age fifteen, making the group one of the small minority with female representation. Obscure Images, the handle of artistic Chicagoland teen Paul Leonard, regularly graced Matt's board, Pure Nihilism, before becoming another mainstay of cDc.

"I'm the pretty much standard-issue, sort-of nerd, moody loner outcast kid," Paul said later. Paul had hung around boards that emphasized trading pirated software, and he was friendly with one of the leading lights of the scene, before the young man became the first person to be tried and convicted under the 1986 hacking law, the Computer Fraud and Abuse Act. After that, Paul was looking for something more fun and more legal. "The cDc people were, at least for the most part, up until the later 1990s, more interested in writing, music, art, and that sort of thing," Paul said. "The technical issues were subsidiary to that." He embraced the do-it-yourself publishing culture that overlapped with music and zines like *Boing Boing*, which morphed from paper to electronic form and is one of the few still around from back in the day. A graphic artist, Paul appreciated and contributed to cDc art made from text characters, which was all that most modems of the day could handle. The group's collective childlike rendering of a dead cow with *X*s for eyes stayed the cDc symbol long after members had the bandwidth to send high-definition movies.

Carrie Campbell provided a lot of the social glue of the group. After a phone call with Warlord to confirm she was one of the rare people with female nicknames who was actually female, he and later the rest of the cDc group welcomed her and treated her with respect. Carrie ran a bulletin board in San Diego and, like the others, phreaked just enough to communicate. She also wrote old-fashioned letters back and forth with Warlord and some of the rest. She never claimed to be a hacker, but she was smart and kind, and the one who kept track of everyone's birthdays.

Except for the Lubbock originals, cDc members rarely met in person before 1990. While their various boards published official cDc

files, they communicated among themselves on a secret part of Demon
Roach Underground. Even Bill seldom showed up in person, because
he went away to boarding school and then to college in Southern California. Warlord finished high school back east, at the private Wood-
berry Forest School in Virginia. With no computer there, he handed off
his board, Tacoland, to Matt. In the summer of 1992, Matt came down
to Lubbock, and he and Bill took a caravan road trip to San Francisco
together, driving separate cars while chatting over walkie-talkies. As
they passed through El Paso, they aimed for the address Matt had for
Warlord's house to surprise him. The neighborhoods got fancier and
fancier, and they finally pulled up in front of a sprawling, stately man-
sion. When a housekeeper opened the door, the boys looked at each
other in disbelief. Warlord had never mentioned that his father was a
well-connected businessman and former county commissioner. "I just
assumed he was middle-class like the rest of us," Matt said. They didn't
have to worry about their lost composure, since it turned out Warlord
wasn't home that day.

Music, especially underground music, brought several members of
the group closer together. Warlord played in a series of minor bands,
while Kevin recorded demo tapes for aspiring musicians and became
a fixture on the local Lubbock music scene. Matt, in Michigan, also
cared deeply about alternative music, which made boards like Kevin's
and Warlord's especially appealing: "In the eighties, it was hard to find
out information about anything that was out of the mainstream." Matt
contributed interviews with post-punk bands Mudhoney and Big Black,
led by future Nirvana producer Steve Albini. Matt went on to form a
small indie record label and publish a music and culture zine, *Cool
Beans*, which took its name from one of Kevin's stock expressions.

Kevin stayed local, attending Texas Tech and working at its radio
station as a DJ. Interested in metal, punk, and rap, he had to stick
mostly to playlists handed down from above. So he faked song requests
from fans in order to spin what he wanted. He played in multiple bands
himself, and in 1995, he went in with Bill and a local skateboard-shop
owner on a venue for live music, Motor 308. He would go through five

locations, rarely taking in more than what he had to pay the bands, before moving to New York in 1999.

While still in college, Kevin took courses in media and advertising. That helped him plot a serious strategy for cDc distribution. A natural "hype man," as he called himself, he would bundle up ten or so new text files and send them out to other bulletin boards for posting. Simply numbering the files was brilliant. That way, bulletin board operators around the country knew if they were missing some, and many would take the time to assemble a complete set. The golden decade of text files would last from 1985 until 1995, when America Online and Netscape's web browser made cumbersome dial-ups to bulletin boards unnecessary. Kevin's strategy, Bill's vision, and the eclectic talents of those who joined them made the Cult of the Dead Cow the best-known and most widely read exemplars of the t-file craft.

[✗ ✗]

Kevin also wanted to learn from hacking's previous generation. A key early find was Chris Tucker, who dialed in from a board in Rhode Island as Nightstalker and became the second person from outside Texas to be asked to join the Cult of the Dead Cow. Chris had gone to Vietnam as a CIA contractor during the war, and he'd come back with a dark view of government power. On his way home in 1971, he read a seminal article in *Esquire*, Ron Rosenbaum's "Secrets of the Little Blue Box." Rosenbaum had spent serious time with the phone phreakers, the forerunners of today's hackers, and he explained what they were doing in plain English. The phreakers were a diverse group, including John Draper, who called himself Cap'n Crunch after learning that whistles given out with that breakfast cereal could be used to blow 2600 hertz, which allowed free calls. The technical puzzles of phreaking would attract future innovators up to and including Apple founders Steve Jobs and Steve Wozniak, who sold blue boxes to make free calls while in college.

The political divide in America at the end of the 1960s was the worst until the 2000s, and that helped push phreaking in a radical

direction. The phone companies were very clearly part of the establishment, and AT&T was a monopoly to boot. That made it a perfect target
for the antiwar left and anyone who thought stealing from some companies was more ethical than stealing from others. In June 1971, Yippie
Abbie Hoffman and early phreaker Al Bell, actually Hoffman acquaintance and former engineering student Alan Fierstein, published the
first issue of the *Youth International Party Line*. The newsletter began
by printing secret calling-card codes and went on to publish explicit
instructions on how to craft blue boxes and other gizmos for making
free calls. After tiring of Hoffman's antics, the publication renamed
itself *TAP*, for "Technological American Party," and continued to make
the most of the free-press provisions in the US Bill of Rights. *TAP* published until 1984, the same year that the essential hacking publication
2600 began to publish.

Chris had his first blue box by January 1972, more than a dozen
years before he connected to Kevin. Chris met another young veteran,
Robert Osband, at a science-fiction convention in Boston in the early
1970s, and the two men bonded over politically tinged stories. Better
known as Cheshire Catalyst, Osband was a ham radio enthusiast and
phreaker, a longtime contributor to the *TAP* newsletter who served as
its final editor. At the *TAP* offices in New York, it was Osband who suggested they host regular gatherings on the first Friday of every month, a
tradition that *2600* later continued in multiple cities. "We were always
about sharing the knowledge," Osband said. "Share the knowledge and
help people build things."

In Lubbock, Kevin had been poring over photocopies of *TAP* newsletters. Now, with Chris, he knew someone who had been part of it.
Chris told stories and patiently answered all Kevin's questions. Chris
had begun computer hacking in 1975, years before *TAP* started covering the topic, and he loved preserving and tinkering with old computers and helping newcomers. Chris lobbied anyone who would listen to
use strong encryption and other privacy tools as they came along, and
he posted purely political cDc files against Reagan conservatives. He
embraced an unregulated internet not only as a great thing but as one

needing active defense in the political realm. Faulting the libertarianism popular with many technologists as a "bottomless pit," Chris wrote
in cDc file "Political Rant #1": "The Computer Underground, once
made up of people interested only in free software, free phone calls,
and flaming each other's hardware, now finds itself having to actually
think about politics and strategies. They have to get involved in the
political process, and they may have to go out and vote, fer chrissakes!"

Kevin wanted to learn what he could from the past so he could plot
the way forward. But the best way to do that was to get together in person, and he was in Lubbock, miles from anywhere.

CHAPTER 3

> THE CONS

THE 1990s STARTED off a lot better than the 1980s for Kevin and the others in cDc. The Berlin Wall had fallen in 1989, George H. W. Bush wasn't as bad as they had feared, and soon Bill Clinton, whom they saw as a reasonable southern Democrat, would take the White House. Computing was still arcane but getting more and more usable, bringing knowledge closer to people everywhere.

Texas had what might seem like a surprisingly strong crop of young hackers. In addition to the arts wing of the hacking community, represented by the earliest members of the Cult of the Dead Cow, there were plenty of others who operated mild-mannered bulletin boards for commentary, community, and, in some cases, conspiracy. On the darker end of the spectrum, some specialized in pirated software and credit cards as well as tips for breaking into big machines at phone companies, corporations, and government agencies. But Texas is a big place, and hackers there had a harder time getting together than their cohorts in New York, Boston, or San Francisco. That kept them from hanging out as much as their peers elsewhere, which meant less fun, less trust, and less deep collaboration and progress.

In 1990, Houston-area hacker Jesse Dryden set out to change that. The proprietor of hacking boards including K0de Ab0de and by then a two-year member of cDc, Dryden was one of a kind: hyperintelligent and deep into music, like some of the others, but possessed of a strong personality both in person and behind a computer keyboard. Dryden came by his passion for music in the most natural way possible: his father was the drummer for Jefferson Airplane, Spencer Dryden, who allied with romantic partner and singer Grace Slick and played a major role in the group's artistic choices. Jesse's mother, Sally Mann, ran off to Los Angeles and then San Francisco in the 1960s. Mann was smart, funny, and so ridiculously good-looking that she could charm her way past any obstacle that stood between her and whatever rock star she was interested in meeting. Her picture was used in a *Rolling Stone* article on groupies, but she was far more than that. She became Slick's closest friend, caught the elder Dryden when he fell from Grace, and in 1971 gave birth to Jesse James Dryden.

Though it earned its own fame, Jefferson Airplane also served as an adjunct of the Grateful Dead, the center of the era's countercul-ture in the Bay Area and by extension America. Grateful Dead gui-tarist Jerry Garcia personally approved Dryden's joining Airplane, and members of both bands and their mutual friends lived together in Haight-Ashbury and other San Francisco neighborhoods. Along with shared creative efforts and antiestablishment attitude, that deep alli-ance meant experimental social structure, early technological adoption, and, as Mann put it, "better living through chemistry." Even before the Dead had their name, they were a part of Ken Kesey's Merry Prank-sters, the eclectic and idealistic group that drove through America to have fun messing with people and to spread the good news about LSD. Another Prankster, visionary writer and marketer Stewart Brand, would also help spread the good news about the coming age of computing. Brand's outlets included the ecology-oriented magazine *Whole Earth Catalog* and the WELL, the pioneering West Coast online community. Among Mann's friends was Dead lyricist and future WELL regular John Perry Barlow. As a Wesleyan college student, Barlow had begun visiting acid guru Timothy Leary, and he introduced the Dead to Leary in 1967. Later, he wrote songs for the Dead, including "Cassidy," a trib-ute to a child that weaves in the history of Beat icon Neal Cassady, still another Prankster. The Dead attended and sometimes performed at Kesey's "Acid Test" parties, and they became technology enthusiasts as well, encouraging the taping of live shows. The swapping of those tapes deepened the Dead's connections with fans and foreshadowed music-sharing services like Napster.

Jesse's unorthodox heritage prepared him well to bring a major innovation to cDc and the broader hacking scene: the modern hacking conference. And it was one of the reasons that Jesse helped turn cDc into a 1990s successor to the Merry Pranksters, as Barlow saw it. Like the Pranksters, the group would exude idealistic joy at tweaking the establishment and describing the rapidly evolving world they saw and that the grown-ups were somehow missing. "Humor is one of the great binding things in the world," Barlow said, and something that cDc shared with the Pranksters was using humor to question the legitimacy of power. As with hackers, Barlow said, "the thing about acidheads is, they think authority is funny."

Though his parents gave Jesse Dryden an amazing start in many ways—intellectual, social, and artistic—stability was another matter. Mann left Dryden and returned with Jesse to Texas from California but spent a brief time in jail. As a twelve-year-old, Jesse talked his way onto the metal band Dokken's tour bus and disappeared for days. Later, he faked going to school for four months. Jesse's computer helped him manage the tension between his shyness and his need for self-expression. "He found some degree of popularity, and he was able to morph that into being out in public and with groups," Mann said. He was a regular at rock clubs, and Jesse also developed an early entrepreneurial flair. He brought in skateboard gear from California and sold it in local parks, then sold rare concert footage. Some of that aspiration took a bad turn, and he was accused of being involved in credit card hacking. "Nothing ever came of it, but they took Jesse's really cool Mac," Mann said.

As the teenaged Jesse's relationship with his mother frayed, he befriended the manager of a local music store, Vince Gutierrez, and lived with him and his daughter off and on. He talked about the Cult of the Dead Cow a lot and introduced friends to Gutierrez by false names or their online handles. Jesse's own came from his description of an LA metal band that he referred to as "drunk fucks." Gradually he became known as Drunkfux or dFx—heavily stylized, with a practiced coolness, and opaque to outsiders. "He has issues of self-esteem," Gutierrez said. "He doesn't feel like he fits in to a certain type of people. cDc was sort of like Jefferson Airplane for him: these cats were just extremely underground. Not in the sense of criminal, but in the sense you didn't know what that world's about unless you're one of them. It's elaborate, like a fraternity."

[x x]

In 1990, at age nineteen, Jesse strategically leaked word on the boards that the "first annual" XmasCon, soon to be known more lyrically as HoHoCon, would convene for three days over Christmas break at a La Quinta Inn near the Houston airport, where single rooms cost $44 a night. The anonymous announcement was short, but it was an apotheosis of Jesse's own style and that of the nascent cDc. It claimed that XmasCon had been planned as a private event before a journalist spilled the beans. Strictly to counter that, he said, XmasCon would be open to the public. And it took several knowing swipes at previous hacking conferences that had been private and that Jesse had enjoyed attending—a three-year-old series known as SummerCon. The first SummerCons were held in St. Louis by the editors of *Phrack*, an online magazine begun in 1985 with a name that merged the words *phreak* and *hack* into something like a curse word. Jesse's announcement ran in an unauthorized revamp of *Phrack* in November 1990.

"We plan on having the biggest gathering of Hackers & Feds since SummerCon '88!," Jesse wrote, inviting "All Hackers, Journalists and Federal Agents." The joke was that while this would be the first hacker conference with feds invited to attend, it was not the first with them present. SummerCon '88 had mainly been about drinking, bragging, and hanging out in real life with people one knew online. But the Secret Service, whose antihacking duties developed from its responsibility to fight counterfeit currency, had shown up and spied on the festivities that year. Nothing all that nefarious was uncovered, but arrests followed anyway. It was part of the buildup to what would be the first law enforcement roundup of hackers across the country, in 1990.

cDc survived those sweeps because it was more of a social space, a refuge for hackers blowing off steam, than a place to plot actual hacks that ran afoul of the law. It also survived the other, related momentous hacking event of that era, the first great battle between two groups, the Legion of Doom and the Masters of Deception. But both developments shaped cDc and ensured its survival. The arrests were a sharp reminder to be cautious where the law was concerned. They also gave rise to the Electronic Frontier Foundation, still the preeminent legal defense group for hackers and researchers, which would intertwine with cDc and its causes. As for the duel between groups, it reinforced cDc's commitment to the pursuit of peace among hacking tribes. In fact, it would have the unusual distinction of admitting members from both LoD and MoD.

LoD began even before cDc, spawned in the early 1980s by a Florida man with the handle Lex Luthor, after the Superman villain. Organization was slipshod, membership was fluid, and regional branches sometimes had little to do with one another. Intriguingly, there was significant overlap between LoD's most impressive hacking adventures and stories in *Phrack*, which grew out of a bulletin board specializing in tales of underground activity. *Phrack* stories circulated on outside bulletin boards the same way cDc files did, but the content included security trade secrets. Unlike the other big hacking publication, *2600*, *Phrack* was online, which left it more vulnerable to prosecution at a time when courts had not explicitly extended freedom of the press to the digital realm. The consequences of the overlap between LoD and *Phrack* would prove important and teach cDc how to stay safe. For *Phrack* consisted of hackers with a publication attached to them, while cDc's file trove would remain a publication first, with hackers attached to it.

HoHoCon's main predecessor was a conference that was smaller and closer to the criminal world. SummerCon gathered just a few dozen *Phrack* contributors and readers to meet in private. The man in charge of the 1988 edition was *Phrack* cofounder Craig Neidorf, who had friends in the Legion of Doom. Attendee Dale Drew of Arizona helped the Secret Service videotape drinking sessions through the wall of his room. That spying was part of a broad effort that culminated in the 1990 arrests of suspects including Neidorf himself. In 1989, Neidorf had published a version of BellSouth's Enhanced 911 manual, an internal document explaining some of how the revamped emergency call system worked. It had been provided by a member of the Atlanta LoD, who was also arrested and pleaded guilty. Neidorf was charged with being part of a scheme to defraud AT&T. By the time of his July 1990 trial, Neidorf was majoring in political science in college and disinclined to settle. Neidorf knew the manual had been stolen, but he hadn't broken into machines himself and had not profited from the theft—*Phrack* was free to readers.

Neidorf's trial became a pivotal moment for hackers and their defenders, in large part because of Jesse Dryden's family friend, John Perry Barlow, the freewheeling Grateful Dead lyricist and early fan of online communities who would be a major influence on cDc. Barlow's fellow acid-taking Deadhead Stewart Brand had spawned the online community the WELL in 1985, and Barlow was a prolific and eloquent contributor. For those with primitive online access via modems, university networks, or other means, it was a mega bulletin board, broken up by topic. Barlow appreciated the dialogue and the chance to connect with interesting people even from his Wyoming ranch.

Barlow's introduction to the rougher side of the internet came in late 1989, when he participated in a WELL group chat about the nature of hacking that was curated by *Harper's* magazine, which printed excerpts. Among those typing in facts and opinions over the course of a week were open-source software crusader Richard Stallman, *2600* editor Eric Corley (under his post-indictment handle, Emmanuel Goldstein), and Cliff Stoll, the Berkeley astronomer who had traced hackers working for Russia and chronicled the work in his book *The Cuckoo's Egg*. Most of the drama came from two brash young New York hackers identifying themselves as Acid Phreak and Phiber Optik.

After Stoll complained that hackers should not be free to enter networks to obtain financial histories from the big credit bureaus, Barlow said he was far more bothered that unaccountable corporations had gathered such data in the first place, which he equated with thievery: "Anybody who wants to inhibit that theft with electronic mischief has my complete support." But after Barlow called Acid a "punk" for lacking vision, Phiber obtained Barlow's credit report and dumped it into the online conversation. "Everyone gets back at someone when he's pissed; so do we," he typed. Barlow later wrote: "I've been in redneck bars wearing shoulder-length curls, police custody while on acid, and Harlem after midnight, but no one had ever put the spook in me quite as Phiber Optik did at that moment."

Even so, Barlow continued to say he was more worried about the government restricting or monitoring computers than he was about the punks. He met the two hackers for Chinese food, reaffirming his belief that they were not the main enemy. Then he convinced Boston software entrepreneur Mitch Kapor, inventor of the modern electronic spreadsheet, and libertarian engineer John Gilmore to join him in founding the Electronic Frontier Foundation. (Gilmore would soon host the Cypherpunks mailing list, which would be home to the most public-spirited cryptographers of the next two decades, along with hackers, assorted freethinkers, and the probable inventor of Bitcoin.) The trio's long-term goal was to extend the freedom of the press, freedom from unreasonable search and seizure, and as many other rights as possible to the digital realm. The short-term goal was to defend hackers who were merely exploring from the full consequences of zealous prosecution, starting with Neidorf.

Until Neidorf's trial, most press coverage of hacking had been full of hand waving and misunderstanding. Reporters were echoing big companies, which preferred to blame their misfortunes on evil geniuses instead of their own poor engineering choices. Reporters were also following the lead of the FBI and Secret Service, where many nontechnical agents and supervisors looking for glory saw greater threats to the world than really existed. But this time, Neidorf had good lawyers, and they showed the court, the press, and the public the major flaws in the case, eventually including the devastating fact that the same information in the manual that BellSouth valued at $79,000 could be openly bought for $13. The government dropped the case, and the EFF was on course to play an outsize role in the debates of the next three decades.

[x x]

Acid and Phiber were actually named Elias Ladopoulos and Mark Abene. Ladopoulos was the first of what grew to more than a dozen members of Masters of Deception, and Abene joined from LoD later. Both were top-notch hackers with a special interest in phone-company computers. At the time that MoD began hacking under its own name, Chris Goggans, a Texan friend of LoD member Scott Chasin, claimed leadership of LoD. (Both Chasin and Goggans would take a turn editing *Phrack*.) MoD taunted the older LoD, and both sides attacked each other in the first big hacker war. It escalated until MoD broke into Tymnet, a system companies used for net connections, to spy on LoD's Goggans and Chasin.

As Jesse started holding HoHoCons, they were natural turf for the Texans in the Legion of Doom, including Jesse's friends, the embattled Goggans and Chasin. As the others at the conference took drugs, drank, and swapped stories, those two hunkered down and plotted. They decided the only way to beat the Masters of Deception was to turn pro. They formed a company, ComSec, and soon convinced Tymnet that its programs were being hacked and that it needed their company's help. Armed with that special access, the pair spied on MoD members and then crossed a red line: they called in the FBI. Ladopoulos and Abene were arrested and prosecuted, each serving a year. But ComSec failed too, in part because the founders' hacking background was too much for the press and customers to ignore. "We were basically blacklisted by the security community," Goggans complained. At one HoHoCon, Goggans told an audience with many admirers that he was angry at how hard it had been. "I'm really pissed. Dealing with presidents of corporations one day, and then you're stuck in a lowlife grunt position trying to scramble for money to feed yourself."

That attitude changed as more companies realized that hackers had knowledge they needed. Chasin went on to found three companies that were acquired by major firms in the security industry and served as a senior executive at number two antivirus company McAfee. Many of their friends gave the LoD men a hard time for going straight and especially for calling the cops. But of those, a large number ended up going straight themselves. "Anybody that could have made a career out of it, did make a career out of it," said Jesse's 1980s housemate Michael Bednarczyk, known online in those days as Arch Angel. "You can start out as 'fuck the man,' but then you become the man, and you start to see things in a different light." Many of the best and brightest in cDc would go straight later, when it was easier. But in general they would avoid the blowback from friends and other hackers that came from working with police and the FBI, going instead to intelligence agencies and the Pentagon.

For all the tension over career paths, gang alliances, and views on law enforcement, HoHoCon was a blast for most people who attended, and it was a major step toward realizing the community that cDc and others had been trying to foster. With informants everywhere, it was hard to build trust, especially online. In person, it was easier. "There were a lot of drugs, a lot of people on acid, but you bond through that," Bednarczyk said. "Now you have someone you've met and trust, and that builds relationships that are pretty strong." In those relationships, people gave information and received it. Everyone learned more about what was doable and how to do it.

Despite the close relationships with LoD leaders, neither Jesse nor cDc took a side in the war with MoD, which ended in the demise of both groups. Among the lessons they took: there was nothing to be gained from battling with peers and breaking the law, and calling in the FBI wasn't wise either. They had the foresight to understand that doing any of these things would hurt you if you ever wanted to do something constructive for the world.

Jesse decided that HoHoCon should be even further above the fray than cDc. He invited not only all manner of hackers but also early professional defenders and even cops, even after a few arrests at the con. "The reason I put on HoHoCon is because I feel it's fairly important to bring everyone from all walks of the computer industry and telecommunications together, both the hackers and phreakers and people from the telco and computer security business," Jesse said. "They can meet each other face-to-face and get good discussions going."

It was easy for cDc to stay neutral because it wasn't a crime board but a place for criminals and everyone else to relax. Not getting involved with crime was a philosophical choice of the founders and early members, informed by the experiences of Neidorf and the rival hacker gangs. But it was also sheer luck: the most technically adept of the three founders had been Brandon Brewer, and he and his older brother Ty happened to have an Atari computer with no hard disk, meaning they could store nothing of any great size—only text files, not programs in any quantity. In any case, the Brewers departed the scene just after it started, leaving cDc in the hands of comparative doves Kevin Wheeler and Bill Brown. "We were like a sperm donor," Ty Brewer said. "We strongly influenced the operation in the beginning and then left. It was our DNA, but that's about it."

[x x]

Another future member of cDc was in LoD, and he was both more of an accomplished criminal and more of a visionary than most who came before. Patrick Kroupa had been in some of the earliest Apple piracy groups before ending up in LoD. In 1991, he founded the pioneering internet service provider MindVox, which was for New York what the WELL was for the San Francisco Bay Area—an early online outpost for people who thought. By Patrick's count, it was only the third place to offer real-time commercial internet connections to the American public. Before then, it had already organized bulletin board–style topics, like the WELL.

Patrick knew Kevin from the 1980s, when Patrick was called Lord Digital and dedicated himself to phreaking to escape his crappy life in Spanish Harlem. He was in it for the challenge of breaking protection around games and then the feeling of control it gave him to dictate to machines around the world. To socialize, from the age of fourteen he went to *TAP* meetings, where mysterious buyers gave him hundred-dollar bills in exchange for hacked information they wanted on various people. Patrick also got on heroin as a teenager, and he stayed hooked through age thirty. He got clean with the aid of a hallucinogen called ibogaine, and he later helped a hundred or so other junkies through the same process, including many fellow hackers.

The Patrick of 1992, still on the hard stuff, amazed people who were coming from vastly different perspectives. He explained MindVox that year in an epic text file that ran in truncated form in *Wired*. In it, he thanked cDc, among others, and gave a personal history of cyberspace to that point that included a frank assessment of many hackers' egotism and criminality, along with admissions of his own opiate binges and unspecified addictions. Patrick said he had eventually realized that the only thing worth doing was helping people. After that realization, he wrote, he reunited with fellow survivors of the busted-up hacker underground who felt the same way.

In 1992, four years before Barlow would pen what would become one of the most famous political documents in the internet's history, his "Declaration of Independence of Cyberspace," Patrick wrote something very similar, an ecumenical and idealistic manifesto. "Cyberspace allows everyone the freedom to coexist without harming anyone else's world-view or belief system," Patrick wrote. He said MindVox would allow users to intersect with pioneers in computer science, the arts, and politics. "Our main priority is to create and continuously evolve an environment that fosters an atmosphere of dynamic creativity, coupled with access to information and ideas, that present you with a far greater spectrum of possibility."

The essay brought articles from New York media, and Patrick gave free MindVox accounts to musicians and artists, marketing the service as a cool one. MindVox remained popular if chaotic for several years, until the Netscape web browser and bare-bones access providers made it hard to charge for sophisticated packages of connections and content. The dawning of the easy-to-use web in 1995 would also be the end of the vast majority of bulletin boards.

In the meantime, Patrick inspired the rest of cDc and helped it stay together. A conference or two a year weren't enough to keep the group whole, and not everyone could get on the group's #cdc Internet Relay Chat channel to keep up with the conversation there. Patrick doled out free email accounts to all in cDc, and Paul Leonard and Carrie Campbell set up an email list to keep everyone in touch.

[x x]

With each passing year, HoHoCon got more prominent speakers, along with more attendees. The second year had one hundred people over three days at Houston's airport Hilton. After a hangover-delayed start to the official proceedings on Saturday, Jesse introduced keynoter Bruce Sterling, the science fiction author whose book on the hacking arrests of 1990 was coming out soon. He plugged the new Austin chapter of Barlow's Electronic Frontier Foundation. The next speakers were LoD hackers-gone-pro Goggans and Chasin, who bragged that five MoD members had been raided earlier in the month. The general debauchery included strippers who gave lap dances to fourteen-year-old boys in the ballroom and turned tricks in the rooms. The hotel managers repeatedly threatened to expel everyone.

At least five from cDc were there, including founder Kevin Wheeler and Matt Kelly, all of whom gathered to live-write some of what would be cDc's two hundredth text file overnight. After slick homages to *Phrack*, teen girl magazine *Tiger Beat*, and the Encyclopedia Brown mysteries for kids, the file described the gogo dancers, drugs, and mayhem of the event, as well as telling an absurdist origin story for the cult that involved monster trucks. Not much of that part made sense, yet file #200 would prove the most popular among cDc's own members.

Other spontaneous meetings helped shape the future of online security. Bednarczyk was walking down the second-floor corridor when a skinny teenager ran toward him. The kid said someone had kicked a plastic beer ball into Sterling's face and the police were on the way, could he duck into Bednarczyk's room to hide? Bednarczyk agreed, and the teen introduced himself as Jeff Moss, the Dark Tangent. Other hackers were already hanging out in the room, and they introduced themselves by handles that were already legendary to Moss. One of them controlled switches at a major phone company. Another had the attack programs called "exploits" that could break into mainframe computers. Moss felt like a mouse in the corner with enormous eyes. "These five people could take over the world if they wanted," he thought. Moss soon would use what he saw, the bonding and the talks and the T-shirts, and found Def Con, the volunteer-powered Las Vegas hacker conference that would become the biggest on the planet.

As HoHoCon grew to hundreds of participants, more new cDc members and future members showed up to meet people they had admired from afar. One newcomer was a talented Boston hacker with the handle White Knight, really named Dan MacMillan. He came to learn and to have fun with old friends and new ones, and like many in cDc, he didn't care who had more underground prestige in the hacking hierarchy. Dan was a pivotal addition to cDc because he brought in more technical people. "We weren't deliberately looking for hacking chops," Kevin said. "It was very much about personality and writing, really. For a long time, the 'test' or evaluation was to write t-files. Everyone was expected to write things. If we were stoked to have more hacker-oriented people, it was because we'd be excited to have a broader range in our t-files." Dan sponsored Bostonian Misha Kubecka, and they would be joined by fellow New Englanders John Lester and Luke Benfey.

As the conferences got bigger, it meant more work and less fun for Jesse. He had *Phrack* or cDc cosponsor the event to share the burden, and then he finally stopped after HoHoCon 5, at the end of 1994. By then, Moss's Def Con had taken what Jesse had started and grown it. Vegas had all the illicit distractions young hackers could want, and the

unruly attendees or the con itself could be banned from any one hotel
and have plenty of others to choose from. If one left drugs sitting out
in the room, cDc learned, the maid would arrange them tidily. Moss
gave cDc free passes and a regular platform, and it would provide some
of the conference's most spectacular moments, drawing more attention
and bigger crowds. A quarter century later, Def Con and its more ex-
pensive spin-off for professionals, Black Hat, would be the dominant
hacker gatherings in the world, attracting the head of the National Security Agency as a keynote speaker and drawing more than twenty-five
thousand attendees to the midsummer desert.

[x x]

Jesse never seemed to find his place. He did some contract programming and cared for his famed musician father near San Francisco,
nursing him through a terminal illness for a year. And he had a series of misfortunes, including a fire that destroyed most of his dad's
memorabilia and a flood that wrecked his stepfather's recording studio.
Jesse took to vanishing for long stretches, leaving his mother and longtime friends in cDc puzzling over where he was. Whatever the problem
was, it was not alcohol or drugs: given his parents' problems, he never
touched either.

If Jesse was unraveling, it might have been because his stories
were no longer holding together. In a young life defined by trauma, he
had found refuge online and with friends who thought he was smart,
cool, and funny, which he was. But he had learned that if the stories
were better, people would think he was even cooler. He told many of
his friends that he played in rock touring bands, with L.A. Guns and
others, yet never sent so much as a cassette to his friend with an indie
record label. He told others he played professional soccer, which struck
some of them as odd given his small stature. He told people a lot of
things, and many of them were not true.

Some of this was about defense—psychological defense, for someone raised in the shadow of famous and successful people, and also

physical defense. Many of Jesse's hacker friends were informants. Telling different stories to different people and mixing the truth with lies kept people from knowing enough to betray him. "At any time, your cyberfriend could become your cyberenemy," Bednarczyk said. "You want to keep your personas separate."

But it was about offense as well. Jesse was a consummate networker, like his mother. He impressed and charmed people and got them to tell him things, and that's how he learned more about hacking, enough to found a critical early series of conferences. Jesse was rude and eloquent, with a rock-star air that made people listen. "He could predict what you were thinking before you said it, then turn a conversation around in seconds," said Angela Dormido, a friend who ran a bulletin board. Hackers have a phrase for the technique: social engineering. It's what made famed hacker Kevin Mitnick so successful, along with many others less well-known. You play a role, you spin lies, and you get people to do what you want. Misha called Jesse on one story that didn't hold up, and Jesse never spoke to him again.

Jesse's mother, two people he lived with at different times, and talented latter-day cDc hackers with resources and connections hunted at length for Jesse after his last sighting in 2009. None admit to knowing where he is, and some who were close to him believe he is dead. They could be right. But in mid-2018, a database showed that he had a valid Texas driver's license, which must be renewed every six years. More plausible is that Jesse used his virtuoso social-engineering skills to fall off the map. Though he might have an excess of that talent, it made Jesse a key part of hacking's development. Like text files, old-school social engineering shrank in importance as technical proficiency grew. As Jesse's time in the spotlight came to an end, the center of gravity in cDc was shifting to Boston, and the group was beginning to move toward bigger things.

> CHAPTER 4

> # UNDERGROUND BOSTON

I N RETROSPECT, IT seems obvious why so many of the attendees of HoHoCon came down from Boston, and why the ultimate college town would provide so much of the Cult of the Dead Cow's new blood. Before most Americans had heard of Silicon Valley, the Route 128 band around Boston was sprouting computer and software companies stocked with graduates from local educational institutions, including Harvard and, especially, its Cambridge rival, MIT. Politicians called it the Massachusetts Miracle. Cambridge itself played host to many innovative technology companies, including two that employed members of cDc and their close collaborators. The better known was Lotus Development Corporation, begun in 1982 by engineer Mitch Kapor. Though Lotus made its first program for Apple computers, it scored a runaway hit with Lotus 1-2-3, the first electronic spreadsheet with graphics. The app worked with early versions of Microsoft operating systems running on IBM personal computers, and it gave many people the first compelling reason to buy a PC. It also earned Kapor enough money to fully fund the Electronic Frontier Foundation, the digital rights group that had saved *Phrack*'s editor from jail.

A few miles away, people who tended to dwell further right on the ideological spectrum were tinkering more quietly. Founded in 1948 by two MIT professors and a former student, BBN Technologies specialized in acoustical engineering before taking on more Pentagon contracts and moving into networking. It helped develop working versions of the internet's basic communication methods, known as TCP/IP, for the Defense Department's predecessor to the internet, as well as early versions of email and other programs that remain classified.

As in Texas, bulletin boards provided the early online gathering places in Boston. Most of those that were open to all comers barred discussion of hacking, making them less attractive to cDc members. The anything-goes exception in Boston area code 617 was the Works, founded by future bulletin board historian Jason Scott Sadofsky. Sadofsky had started the Works during high school in Chappaqua, New York, in 1988. He handed it off to a user to run two years later when he moved to Boston for college, where he would preside over the scene starting as an old man of nineteen. The Works ran cDc files, naturally. And it was a gateway to more serious hacking boards. On the Works, the adventurous could find mention of closed boards hosting discourse that was riskier, or that would have been if the proprietors had not closely vetted the participants to screen out cops, snitches, and the overly talkative. Invite-only boards where discussions could wander into legal gray areas included Black Crawling Systems, Calvary, and one called Democracy, which morphed into ATDT, after a modem command for dialing a call. This last was run by roommates who used the handles Magic Man and Darby Crash. In 1991, Boston University grad Darby left town for a job with Microsoft. Under the name J Allard, he would push Bill Gates to put internet functionality in Windows 95 and later run the company's Xbox division.

John Lester had been on bulletin boards for years from his childhood home in Dartmouth, Massachusetts. He attended MIT and afterward went back for more online fun under the handle Count Zero, from the William Gibson novel. While working at Harvard-affiliated Massachusetts General Hospital on Alzheimer's research, he wrote technology explainers for both *2600* and *Phrack*. Magic Man made John a co-sysop of ATDT when Darby Crash moved west, and John inherited the whole thing when Magic Man left for Colorado. Brian Hassick's Black Crawling Systems was highly technical. Calvary, run by Golgo13, had a tougher crowd: Golgo13 liked to break programs. The login sequence featured a picture of Jesus on the cross and a slogan: "You bring the hammer, we've got the nails."

One day in August 1991 brought everyone together, prefiguring what would become known as the L0pht, the first shared hacker space in the country and a powerful symbol of hacking's positive potential. John and Darby lived in the same apartment building near Fenway Park, home of the Red Sox. They decided to host a barbecue on their roof to get the regulars on their various boards together, calling it the Grill-a-Thon. Everyone was instructed to bring their own food to cook. For the majority, it was the first time people had ever set eyes on each other, even if they had chatted online for years. It was then John met the mysterious Golgo13, who made quite an impression: Most of the kids were skinny, nerdy, and pale. Golgo13 was a big man who arrived riding a badass motorcycle, looking exactly like the rock-club bouncer they later discovered him to be. Then six-foot-six Luke Benfey, known as Deth Vegetable, showed up and towered over Golgo13, though he was much milder in manner. Soon-to-be cDc members Dan MacMillan and baby-faced Misha Kubecka were also there, and a fourteen-year-old kid with the attitude of a delinquent, Joe Grand, known as Kingpin.

A football game broke out, and Luke carried the ball on a play during which Golgo13 seemed determined to stand out. He launched a flying tackle at Luke as another hacker did the same from the other side, shielded from Golgo13's view by Luke's enormous bulk. Luke went down, and Golgo13 got up with a gash over his eyebrow that bled so profusely into his eye that he couldn't cycle home. Instead, he walked a few blocks to Beth Israel Hospital for stitches and came back for more beer. Grand, in the meantime, had been kicking coal off the roof at people walking below, one of whom had called the cops. The elevator was out of order, so by the time Boston's finest made it to the roof, they were out of breath and extra mad. John and the others apologized and plied them with sausages. They accepted, but warned: "If we have to come back here again, somebody is going to jail."

Even without the blood and the cops, it would have been a memorable day. Finding out what people were really like cemented relationships that would last decades; the annual Grill-a-Thons themselves are still going as of 2018.

[x x]

Soon John moved into Hassick's building in the South End. Both of their girlfriends complained about the computers and other odd equipment spilling all over their apartments, much of it bought cheap at the MIT flea market for discarded electronics. The two women were trying to start up a business of their own, sewing decorations for hats, and there simply wasn't enough apartment space for both projects. In 1992, John found an artist's loft with cracked floors and character a stone's throw away, on Waltham Street, and all four started using it for their hobbies. It was a loft, but when they referred to it in writing, John called it something tongue-in-cheek: the L0pht, with a zero instead of an *O* and with the "ph" from *phreaking*. It was leet speak, the joking "elite" language of hackers. John and Hassick then rented out desks to their friends, including Golgo13, Dan, and Grand, who was brought in as a way of keeping him from following a criminal path. Grand preferred messing with hardware gadgets to software, putting him well ahead of the chip security and maker movements to come. But he had not shed enough of his punk attitude when he met the older hackers. He was getting access to credit bureaus with stolen passwords, pulling information on doctors and dentists, and then calling banks and asking for new credit cards in those names. The turning point came in 1992, after he broke into a Michigan AT&T office, avoiding jail only because he was a juvenile. Grand's parents let him keep hanging out at the L0pht, realizing the older guys there could help redirect him, and they did.

When the hat business failed and the women moved their stuff out of the L0pht, it made room for a few others. "Brian and I had this vision of it being kind of a clubhouse anarchic learning lab where people could bring hardware and take it apart. We could leverage each other's expertise as well as existing hardware," John said. "People who had a lot of potential in certain areas could meet people who could maybe mentor them and introduce them to others." John and Hassick had just founded the first enduring hacker space in America. For the next eight years, the L0pht would be one of the great hot spots in hacker history. It would host cDc's first website and eventually share four members with cDc in a kind of coevolution. Admirers founded similar spaces around the country. John saw it as a sort of 3-D bulletin board, a permanent bridge between the digital and physical worlds: "a communal clubhouse / think tank / meeting place / storage place for hardware / communal library" and crash pad.

John had read cDc files on the Works, and he joined Misha and Dan in a Boston delegation to HoHoCon at the end of that year, 1992. He ended up in what he dubbed the "Suite of the Elite," the biggest and most communal hotel room at the con, which would become a standard feature of any con with a cDc contingent. There too was Kevin Wheeler, Swamp Rat himself. It was late, they were tired, and they chatted about a number of things. Then, John asked casually, "How do you get to be in cDc, anyway?" Kevin explained that since he started it, "it's just basically if I say you're in." Oh, said John. "Could I be in?" Missing the point and responding theoretically, Kevin said: "Yes, you could be in." Rolling his eyes, John kept going. "May I be in?" And Kevin waived a rule against letting in people who asked to join. "Okay, sure. You're in the Cult of the Dead Cow."

[x x]

Despite the overlap, there were important differences between cDc and the L0pht. The former had no physical place and no rent to pay, and it included a greater variety of people. The lack of an address also made it easier for cDc to stay darker and more mysterious, and more easily associated with the criminal underground, especially when it chose to play that up. But in truth, the L0pht also attracted people with a range of attitudes toward activity that approached or crossed legal lines. There was no one whose chief goal was hacking for profit, but that still left a lot of room for varied approaches. John Lester admits to having used pilfered calling codes, as did pretty much everyone else, to dial into boards long distance as a teen. His best friend and partner in L0pht's founding, Brian Hassick, said he also bought things with stolen credit cards. And Golgo13 said ATDT, which moved into the L0pht with John, was "an actual den of hackers," including some who discussed "carding," as dealing with stolen credit cards was called. On the closed boards, people would share "dial-outs," codes to punch in from inside a local business's phone network to make free long-distance calls. "I am not [in] the noble pursuit of making something better, hacking as a way to learn more," Golgo13 said. "I hack things because I like to screw with stuff."

Like the much younger Joe Grand, Hassick was trying to move on from a dark history under the handle Brian Oblivion, taken from the movie *Videodrome*. The son of a Pennsylvania steelworker and a go-go dancer, Hassick had tapped into a neighbor's phone line so he could "war-dial" numbers from automated modems on two phone lines simultaneously, to see who or what would accept the connection. Hassick got into heating and other systems and once turned off the lights at a mall. He left home at fifteen but stayed in his high school through graduation, when he hopped trains to Seattle for a change of scene. He came back east in 1989. Despite having decent technical skills, he took a job working the overnight shift at a convenience store in Charlestown, the tough Irish neighborhood in Boston shown in movies like *The Departed*. His store was robbed on his shift twelve times. Hassick was familiar enough with the rules of the street. He gave none of the robbers any trouble.

Hassick and others who would power the L0pht and cDc were born in the period 1969–1971. That made them the perfect age to take advantage of a magic window between when *War Games* came out, in 1983, and when the Computer Fraud and Abuse Act made unauthorized computer access a criminal act, in 1986. On average, kids born in those years were also more likely to have young parents with a critical view of the US government. Dan MacMillan, the first Bostonian in cDc, was born in 1969, and he epitomized both factors. His father, from blue-collar Cambridge neighbor Somerville, had plenty of friends in the Irish Winter Hill Gang. To avoid a similar fate, MacMillan's father enlisted in the navy, learning Morse code and cryptography as an entry-level intelligence officer. That led to a CIA analysis job. He saw too much bureaucratic politics inside the agency, grew disillusioned, and quit, preferring to work for himself as a mechanic than to be part of a giant amoral machine.

Dan grew up an independent thinker in Brockton, the same working-class Boston suburb that would produce Napster founder Shawn Fanning. His father didn't mind spending money for his kid's computers. Dan had something of an offline life, running track and playing volleyball, but he spent time on bulletin boards and learned enough to get paid for setting up databases for local businesses as a high school sophomore. He quit school to earn an equivalency certificate, and technical courses at a college in Vermont didn't hold his attention either. Dan's questionable digital activities before leaving high school included shutting off the school's heat in the dead of winter on a day he didn't want to go to class. He also obtained some computer equipment he couldn't pay for and used red boxes for free phone calls from phone booths. Later, with soft-spoken California transplant Misha Kubecka and others from ATDT, Dan breached various institutions to learn what he could.

With still-developing laws, poor corporate defenses, and few role models beyond Chris Tucker (Nightstalker) and others with antiestablishment Yippie leanings, people drew their own moral lines. Dan said he wouldn't read others' email. And, like Hassick, he cared about privacy as a broad social issue, enough so that he and Misha wrote a 1992 text file for *Phrack* pointing out all the poor controls at a big data broker of the day, Information America. But in addition to faulting the poor security, the article gave strong hints for hackers who might want to research individuals. Among other things, it noted that "initial passwords, which are assigned when an account is first created, are usually composed of the account holder's first name, or first name plus a middle or last initial." Later, Dan would regret being so explicit. Even after the file's publication, he continued to have easy access to the data broker. Once, he used its address database to help his uncle deliver a large number of toilets to someone who had wronged him. Another time, Dan looked up personal information on an actress he thought he might be able to date, but he said he never used the data.

These old-school, semi-public-spirited hackers didn't like stalkers, professional criminals, or informers like Agent Steal, who had gone to SummerCon and secretly taped cDc members, but failed to catch them confessing to crimes. Steal did turn in Kevin Mitnick, future *Wired* journalist Kevin Poulsen, and others. All of that "cheapened the scene," Dan said. "The conceptual stuff in security is more interesting than helping bust people." He had gotten to know Kevin Wheeler on Demon Roach Underground and again on hacked conference calls. Alliance Teleconferencing was a favorite target. With a hacked account, Dan and others would avoid calling fees by setting up conference lines that were free to call in to for days or weeks. Sometimes only friends and allies were invited. At other times, for fun, the organizers would keep it interesting by conferencing in talk-radio personalities, crazy people, and phone-sex girls.

After Kevin inducted Dan into cDc in 1990, Dan sponsored fellow Bostonian Misha Kubecka, known as Omega. Misha wrote well and took up editing duties on cDc text files, helping set the overall tone. Like others, Misha had followed the credo laid out by early hacker the Mentor, who urged exploration and not destruction. Later, upset by how much personal data was collected by Information America, Misha got very serious about individual privacy while still believing that technical information should be shared: "You could get anything on anyone. It was a shock to White Knight and me and others, and from that moment on, privacy was extremely important for me."

The last full addition to the L0pht's first location was Chris Wysopal, who had picked a spot on a Massachusetts map without looking to get a handle that no one else would have: Weld Pond. All the desks were spoken for by then, at $200-per-month rent. So he split a spot with Joe Grand, each paying $100. Chris had grown up more conventionally than the others, and he came across as less rebellious. The son of a General Electric engineer, Wysopal attended a Catholic high school on the North Shore outside Boston, then went to Rensselaer Polytechnic Institute in Troy, New York, which ranked in quality behind only MIT and CalTech in many computer degrees. At RPI he hosted a hacking bulletin board that attracted some from the Legion of Doom, but he didn't get in much trouble himself. Returning to the Boston area in 1987, Wysopal got a coveted job at Mitch Kapor's Lotus Development and stayed focused on that. But a few years later, he started hunting for bulletin boards again, landing at the Works and Hassick's hard-core hacking bulletin board Black Crawling Systems. A few months later, Hassick invited Wysopal to the L0pht as well.

Now including John, Hassick, Golgo13, Dan, Grand, and Wysopal, the L0pht crew would go "trashing," diving in dumpsters outside phone company central offices or corporate buildings. They were not looking for the carbons of credit card slips, known as "black gold." They wanted usable equipment, and manuals, and perhaps an internal phone directory—anything that would list what machines and software were running inside and hint at how to get connected and operate once in. But they kept shopping at the MIT flea market too. As much as possible, they wanted their hacking to be on the right side of the law, tinkering with what they owned themselves. "That was the genius of the L0pht that took a while for people to understand," Wysopal said. "We could learn on our own computers and not have to steal anything." Staying clean was especially important as the group grew more public with its research, which was generally alarming, since the state of security was appalling. Once, when the group had discovered a vulnerability in Microsoft software, a visiting reporter was confused. "You mean you can break into Microsoft with this?" Well, yes, Wysopal told her. "But you can break into any computer in the world with it."

[x x]

The annual Grill-a-Thons continued and spread to the West Coast. But there were other events that sprang up more often. Sadofsky's the Works began holding small monthly meet-ups in Harvard Square in January 1991 at the urging of Misha. Those soon expanded under John Lester into the area's *2600* meetings. The gatherings started in Café Aventura, on the second floor of an indoor shopping center called the Garage. When the weather was nice, they often moved to the outdoor tables at Au Bon Pain, across the street from Harvard Yard. Later, when too many people came, the first-Friday meetings moved to the Prudential Center in downtown Boston. It was an unstructured show-and-tell and social hour, with people moving from table to table. After the meetings, smaller groups would head into the Square or to MIT, where they could monkey around with pay phones, explore the tunnels, or abuse the internet terminals in the lab. MIT was home to open-source fanatic Richard Stallman, who didn't believe in passwords, and the same ethos contributed to what would otherwise have to be seen as very poor security practices. Among them was the lightly guarded secret that any lab terminals would grant internet access to the username "root" and the password "mrroot," later upgraded to "drroot." Often enough, old-timers would finish the night at Sadofsky's apartment. It was on one of those occasions that Misha and Dan MacMillan realized that they had known each other for two years online.

Many in the Works and *2600* crowds were teenagers. One, Limor Fried, began coming as early as age twelve. Known as Lady Ada, she would go on to be a pioneer of the maker movement and the first female engineer to be featured on the cover of *Wired*, helping educate and inspire with Adafruit Industries. For those handling information as sensitive as unpublished software flaws, twelve or thirteen was too young to trust. So the more experienced hackers would wait until the *2600* meeting wound down and then head off to a nearby bar for what they called 2621—the meeting of the subset old enough to be served alcohol. Only then would they bring out the printouts of the holes they had found and pass them around. The one deemed best earned its maker free drinks.

"You didn't tell anyone. It was like *Fight Club*," said participant Jordan Ritter, who belonged to a hacking group called w00w00 and designed the server architecture at Napster for fellow w00w00 member Shawn Fanning.

Even without admittance to Sadofsky's place or 2621, the monthly meet-ups were a great place to find out about other boards, plan road trips to cons, and hunt for roommates. One of the most noticeable underage regulars was the enormous and exuberant Deth Vegetable, who would become a cDc leader. Born in 1973 and raised in a succession of New England towns, Luke Benfey had managed to talk himself into a slot as co-sysop of the Works, and he seemed to be curious about practically everything. It was a continuation of the liberation he had first felt online. He had been playing with computers from age seven, something made possible by his father's job at big VAX manufacturer Digital Equipment Corporation. Despite the establishment job and a previous Air Force stint, Luke's father was an old leftist and self-described beatnik, a Holocaust survivor who had come to America as a teenager. He was therefore preconditioned to be flexible about his son's confrontations with authority, which began not long after a cousin showed him *Phrack*. In 1987, Luke's parents got a $600 phone bill and there was an unpleasant conversation. Like virtually all of his future friends, Luke found other and less legal means to connect. The magic of the early internet meant that other people had grappled with the same issue, figured out what to do, and written text-file tutorials. Luke consumed those, other takes on technology, and anything edgy or funny. By fifteen, he was copying what he thought might be interesting to his own fledgling board, including a grab bag of anarchist files with pipe-bomb instructions.

Luke became a fan of cDc after reading its files on the Works. cDc people had skills but didn't take themselves seriously: they were an enormous inside joke for hackers. Any industry has its own leading figures, language, and perhaps even running gags. But hackers were especially misunderstood by outsiders, so many bonded by complaining about the misconceptions, incomprehension, and stupidity. cDc managed to make fun of both more self-aggrandizing hackers and the clueless public, making it seem effortless. That was cool.

Luke did a bit of hacking on his own, including trolling around with a bug in the email program Sendmail. In early 1991, he grabbed some file directories from a US military base in Subic Bay in the Philippines, just to see what it was foolish enough to leave accessible. He saw what looked like notes from a Defense Intelligence Agency briefing that described a coming invasion to retake Kuwait from the Iraqis, including names of units that would be involved. After the airstrikes began, Luke realized that he had been looking at the real thing, not just one of many scenarios. Even though he opposed the war, he realized that distributing the plans might mean espionage charges.

With Misha and others vouching for him, cDc took Luke in the next year, and he made his pilgrimage to HoHoCon in 1993, the first time he could afford the trip. "White Knight and Misha and Golgo13 had gone to SummerCon and previous HoHoCons and come back with these amazing stories," Luke said. "It was dark and mysterious," a conference for people who probably shouldn't have conferences. When he got there, Luke tried not to come off too awestruck when hanging out with Jesse and Kevin, who sported reddish blond hair to his midchest, controlled by a cDc-branded baseball cap. "I was part of cDc, but they had been doing it for years, and they were guys I looked up to."

The living situation in Boston was fluid. In 1993, Luke moved to a place dubbed Messiah Village, sharing space with a group of hackers and goths and oddballs, including future cDc member Sam Anthony, known online as Tweety Fish. Sam got some social conscience from his mother, Amy, an expert in preserving affordable apartments who served as the top state housing executive under Governor Mike Dukakis. Sam was even younger than Luke; born in 1975, he didn't get a modem until 1989. But he was a fast learner, making it to the Works meet-ups by the following year.

One day at Messiah Village, a crew from the NBC News show *Dateline* appeared. In 1988, at fifteen, Luke had written a text file that combined a pipe-bomb formula with doggerel about slimming down by losing limbs, producing a piece like Swamp Rat's gerbil feed file. A board operator in Connecticut copied it. The police had their eyes on that man, and after a fourteen-year-old downloaded it, they busted the operator. News of that bust sparked interest in Luke's file. Kids searched for it, including three teenagers in Montreal who injured themselves in two pipe-bomb incidents. One lost parts of two fingers. A spate of such occurrences got major press as bulletin boards grew in popularity and parents realized their children were getting access to anarchy files and pornography.

Quite sensibly, most involved with sketchy boards whom *Dateline* contacted declined to talk. But Luke thought that the issues should have broader debate and that it would be fun to be on television. When the episode aired in September 1994, Luke said he was devastated that kids had hurt themselves, explained that the file was a joke, and argued reasonably against government censorship. *Dateline* explained that Luke's handle was Deth Vegetable. The exposure and resulting hand-wringing by outraged politicians, of course, did nothing but tell more teenagers where to look for sketchy material.

A second hacker haven on Mission Hill was nicknamed Hell: it housed future cDc electronics whiz Charlie Rhodes, known as Chuk E, and long-haired San Franciscan Dylan Shea, called FreqOut, who would also join cDc. Dylan had moved up from his second hometown in Madison, Connecticut, and felt lucky to have fallen in with the *2600* crowd. Someone he met at one of the get-togethers taught him how to make a red box for calling anywhere from a pay phone. He and Charlie were enrolled at nearby Wentworth Institute of Technology and had access to a lab where they made circuit boards to mass-produce the devices, selling them to other students for $30 or $50, just enough to afford more gear. They would have felt uncomfortable going for bigger profits, and they took pains to avoid selling to drug dealers, a natural market but an unpleasant one. Poetically enough, Hell caught fire after a suspected arson attack on a nearby triple-decker.

In 1995, the two living groups combined in Allston at a place dubbed New Hack City. It housed Luke, Dylan, Charlie, and Window

Snyder, known online as Rosie the Riveter. The Choate-educated daughter of two software engineers, one an immigrant from Kenya, Snyder was analytical, intense, and sardonic, but kind. She was also a fairly rare sight in American hacking circles back then as a black woman. Snyder would go on to play major security roles at both Microsoft and Apple. "That place [New Hack City] was knee-deep in Taco Bell wrappers," Snyder said. "It was the most disgusting place I ever lived, but also the most fun I ever had."

The Nielsen television ratings company picked the house for one of its devices, and the group fittingly decided to use its outsize influence for good. The only TV set that the Nielsen people thought was there stayed tuned to the public station constantly, except when a visiting hacker wanted to give another favorite show a boost. Snyder didn't stay for long, because a boastful hacker named u4ea breached a Pittsfield internet service provider and threatened to do much more. In the ensuing local media frenzy, the *Boston Herald* identified New Hack City as one of five major Boston hacking groups, adding that its members had been interviewed by police. A resident close to Snyder had been arrested as a teenager and wanted no more of the wrong kind of attention, so they both took off.

The internet, and Microsoft, were about to be everywhere. Netscape, the first browser, made what there was of the web easy to surf. But the mass inflection point came in August 1995, when television talk show host Jay Leno joined Microsoft's Bill Gates to launch Windows 95 in a media spectacle that would become commonplace for consumer technology releases. The TV commercials were everywhere, playing the last great song by the Rolling Stones, "Start Me Up." The newspapers and magazines were full of giddy explanations. Everyone's grandmother now knew about getting online with a computer. Unfortunately, no one was saying she needed to be careful doing so.

As the Cult of the Dead Cow's technological sophistication had ramped up, its social sophistication now had to ratchet up as well. Not everyone in the Boston scene had serious white-collar tech jobs, but more began to get them as the broad public internet arrived and launched an unprecedented technology boom. Yet many of them had dabbled in crime, and pretty much all of them were friendly with people who had been or were still regularly on the wrong side of the law. To be accepted and admitted by both the hacker world and the straight world was like walking a tightrope over a minefield.

Your hacker buddies wanted you to bring home source code, the programmer's work product, for a "security audit," just to be sure you hadn't botched it or sold out. Your current or future employer wanted you to have experience, but it couldn't be seen to know too much about how you got that experience. And nobody of any background liked a rat, except the FBI, which was the only element able to put you in jail if you didn't say what you knew about your friends.

In the 1990s, there was one person who proved able to completely master the worlds of semicriminal hackers, straight security, and the government to boot. The best known of his names is Mudge.

> CHAPTER 5

> BACK ORIFICE

PEITER "MUDGE" ZATKO arrived at Boston's Berklee College of Music in 1988 to study guitar composition and performance. It was either that or go study technology somewhere else, and back then computer science departments weren't teaching what he was interested in—how things really worked, as opposed to how they were supposed to work. But his classes during the day were not going to present much of an obstacle to learning what he wanted. Mudge already knew a great deal from experimenting and from the bulletin boards he had been on for years, where he had met Dan MacMillan and others. Once he moved up to Boston from his father's place in Pennsylvania, Mudge also found the *2600* gatherings and discovered that MIT students were just as interested in using Berklee's recording studios as he was in using MIT's lab computers. Bartering solved both problems neatly.

Mudge stood out in many respects, even from oddball hackers. He grew up in the deepest South, where his father, David, taught sophisticated chemistry at the University of Alabama, and he was a full-fledged musical prodigy. His parents started him off carrying a cigar box under his chin at two and a half, Mudge said, to get him used to putting a violin there. By the time he got to Berklee he was practicing five hours a day, a routine he compared to the grueling training of Chinese acrobats. But he was never just about music. David Zatko worked on the government's space shuttle efforts and brought home computer parts to his toddler.

With a $5,000 bequest from Mudge's grandfather, the middle-class family bought an Apple II Plus, intending it to be educational. That it was, especially because a nearby store offered software that the buyer could return quickly for a partial refund. That made cracking the copy protection an imperative for Mudge and his father, and it was an early lesson in perverse incentives, a subject that Mudge would one day find himself debating in the Pentagon. Breaking the rights management on Apple software and games like *Ultima IV* "was our jigsaw puzzle," Mudge said. "We did that, and we picked locks."

Before the Computer Fraud and Abuse Act of 1986, and especially before *War Games* made open networks into overcrowded playgrounds, Mudge roamed far from home. His custom when entering a company's network was to leave a message announcing himself. Sometimes, the administrator would bark at him to leave. Other times, employees would ask him to avoid a certain area. But most often, no one complained. Given Mudge's attitude, his skills, and the LoD and MoD members he hung around with, many of his friends believe Mudge did other things that would be harder to defend in the light of day. Officially, he denies having broken the law, even by uploading pirated software to the trading sites he visited. He admits only that he got unwanted attention from the authorities due to his explorations. Others who might know differently could have a tough time proving it was really Mudge they were dealing with. When it came time to fill out forms to apply for a US government security clearance, Mudge's list of aliases ran for ten pages.

Obviously Mudge had been up to something—so much so, he joked, that when the Chinese stole his and millions of other people's SF-86 security-clearance applications in 2015, they must have thought they were being trolled: no one with his history could have gotten a clearance. To remind him of the risks of overstepping, Mudge kept a picture above his computers of his friend Byron York, known as Lou Cipher, getting arrested in Texas. Mudge had met York through Dan MacMillan and Jesse Dryden. To Mudge, he was a nice guy who had been through a lot. In the picture, York was face down on the grass, a cop's knee on his back.

The picture also served to prod Mudge about discretion. At Ho-HoCon '92, when he was out on bail, York had told his fellow hackers that he had been set up by a full-time informant who preyed on his circle after one of them admitted to crimes during a meeting of Alcoholics Anonymous. "He badgered us for about six months until we finally said okay, allegedly" to a scheme counterfeiting government checks, York had said. "Entrapment doesn't apply because he's not law enforcement."

The snitch was in the background of the arrest photograph, unmolested. The picture changed offices every time Mudge did, "a constant reminder to never lose track of my moral compass and why I was doing everything, and that it would require constant vigilance to do so," Mudge said. He developed his own ethical code: He cared about information. He didn't care who he got it from, including criminals, and he was generous about sharing it, including with government officials. But he would never name names.

After Mudge moved to Pennsylvania as a child, his parents' bitter divorce left him in control of his own hours. He convinced his suburban public high school that he was an emancipated minor and could excuse his own absences. Mudge preferred to hang around older musicians and hackers, including Robert Osband and others he met through *TAP*.

Then came Boston, and meeting his fellow hackers in Harvard Square, and after college a trainee slot at BBN Technologies, working with people who helped build the internet. The long-haired Mudge started out in a temporary tech-support job in the supercomputer department, with the promise that he could stay on in another division if it agreed to take him. Instead, he signed on to create the security department. By that point, he had already gotten Dan a job at a different computer company. Over the coming years, he would help Brian Hassick, Chris Wysopal, and several others land jobs at BBN.

Dan took Mudge to visit the L0pht in 1994, and two years later, as the consumer internet was taking over the outside world, he joined. Around that time, the group was moving to a bigger space in a warehouse in Watertown. Mudge immediately started spinning ideas about making the L0pht more sustainable. Instead of just a clubhouse, he thought, it could be a research lab. They could make security tools and sell them, using the money to keep hacking. Eventually, if all went well, they could quit their day jobs and hack whatever struck their fancy.

There was one hitch, the existing members said: cDc's John Lester, known as Count Zero. He wasn't interested in turning their hobby into a business, and he felt it would fundamentally change the chemistry of the L0pht. One night, while everyone was together but John, they sent a cowardly email from L0pht cofounder Hassick's account asking John not to join them in Watertown. At a follow-up dinner with him to discuss it, Mudge did most of the hard talking. His role in John's departure cemented Mudge's new position as L0pht front man.

Newly incorporated as L0pht Heavy Industries, the group began releasing tools, including one that originated at Mudge's day job. He was used to Unix, but BBN was bringing in Windows machines, and Mudge had to handle security on those as well. Looking to test the strength of user passwords, he discovered that Microsoft was chopping up long and strong passwords into two fields of seven characters each, making them far easier to crack. He wrote a guessing tool and asked BBN if they wanted to do anything with it, but the program had a casual, homemade feel to it, and BBN declined. So Mudge brought it home to the L0pht, which put it out as L0phtCrack. Wysopal wrote a second version, adding a graphic user interface, and the L0pht began charging a small amount for it.

The L0pht also released a series of security advisories, warning the public of flaws in a range of software, including Sendmail, Lotus Domino, and Microsoft's IIS web server. Security consultants took note and customers complained, forcing the product makers to issue fixes. The advisories drew the first wide attention to the L0pht. And within the industry, it crystallized a debate that had been raging behind closed doors for years. Many companies argued it was irresponsible to tell people about flaws in the private software they sold because it taught hackers how to break into customers' machines. In some cases, software producers even sued researchers for evading the protections on programs they had purchased in order to look inside. But when hackers told only companies of the flaws, the software makers usually ignored them. The only way to actually force things to get fixed was to expose the information.

[x x]

Given what Mudge had accomplished at the L0pht, Misha Kubecka and Dan MacMillan lobbied Kevin to bring Mudge into cDc as well in 1996. "Mudge is someone to be reckoned with, and it's a good idea to have him in our camp," Misha wrote to the group mailing list. The others had the impression that, among his other qualifications, Mudge had hacked other security luminaries. But Mudge generally let people think he did more hands-on hacking than he did. At BBN, he had free rein over everything that company supported, including military and financial systems. That made random break-ins elsewhere less tempting. Once, a leading security figure came to the L0pht, and Mudge asked him why the White House email-monitoring system the visitor had built had been configured in a certain way. Mid-answer, the guest realized that Mudge had to have been inside that system to know enough to ask the question, and he said as much. Others present assumed that Mudge had hacked the White House, though actually he had been authorized to examine the design on behalf of BBN.

At the L0pht, Mudge also acted as a defender. He did install a back door on the Unix servers to make sure they weren't misused, or at least not much, by guests. But outside his home turf, different rules applied. Mudge wrote exploits and gave them away to defenders and attackers alike. "I would give certain teams, groups, and people early access to some of my software and tools. Sometimes tools that were a bit too powerful and purpose-built for me to release them publicly," Mudge said. Sometimes, those attackers would give him back tribute, including priceless code for major operating systems. Mudge did not ask for those goods or trade for them, and though in theory he could still have been charged with receiving stolen goods, he was not.

"The bartering system back then for actual hackers and folk were these tarballs of proprietary source code, personal or private information. New tools were sometimes viewed as more valuable, so I was looked at as a real heavyweight," Mudge said. "It was important for me to be viewed as sharing with the community, because I believed in it. And yes, there are parts of the community that were obviously doing illegal things. That wasn't my focus, nor my goal. I wanted to inspire more people to release novel tools and applied research so we could understand and fix the cyberworld that was being erected around us."

Though there was enthusiasm for Mudge from the Bostonians, Jesse, and others, Kevin had final say over all new members, and it was going to be awkward because John Lester was already a member. But Mudge would cement the group's transformation from self-publishing pranksters to actual authorities on security. Kevin made the deal.

Mudge got something from the marriage as well. He wanted to "make a dent in the universe," he said. A hacker's hacker, he wanted to tear things apart and find out how things really functioned, then either explain them or, if possible, put them back together better than they had been. He applied the same mind-set to other aspects of the world—the computer industry as a whole, politics, and the media. The mainstream media was evolving as the web gave so many others a voice, but it was still a dominant and mysterious force in the world. How did it decide what was true, and which truths were more important? How did other factors come into play, like the sex appeal of a story, potential audience size, and the pursuit of the greater good?

cDc had been moving into a phase of "culture jamming," playing with the media, as the group became better known. Mysterious criminals messing with not just strangers' home computers but NORAD's mainframe made for great copy, and cDc had decided to help explain things, at whatever level the reporter was at. If reporters asked serious questions, they would get serious answers. If a clueless TV correspondent just wanted to hype something as scary, cDc would accommodate that too. The group realized that coverage led to more coverage, especially when so many knew so little about computers. "In the right vacuum a group like cDc can flourish. That's their talent," said the Works founder Jason Sadofsky. Kevin, the self-described hype man, had been thinking about the distribution of text files when most people were not. Now there were cameras showing up, and cDc had some credibility, and they ran to the cameras, Sadofsky said: "Here we are! We're hackers!"

Mudge saw a chance to learn. "The experiment was, how easy is it to manipulate the press and the media, and this is actually fairly relevant right now," Mudge said in 2018. "If we say something, will it actually be repeated? They would jam information to see how far out it would go. I thought it was fascinating. It made me look at the media in a different way. I started to understand the incentive structures and the restraints on resources." Mudge took what he learned and applied it back at the L0pht, which shared some members with cDc and was working on similar problems, but which was treated more respectfully by reporters and TV crews. He got to play both good cop and bad cop in the security world.

[x x]

Though the arrival of the ubiquitous web in 1995 killed off most bulletin boards, cDc managed the transition because of its expanding cast of actual security experts and its physical base at the L0pht. Just surviving was half the battle. Once it did that, cDc's history made people turn to it when they wanted to know where internet culture was coming from, what the web meant, and how secure it all was. Those who stumbled onto cDc then touted it to others. It was a real resource, but it was also an inside joke turned pro.

The media, of course, were hardest-pressed to explain the web, and they came early and often. When they searched, pre-Google, for news about hackers, they would find Luke Benfey's 1994 *Dateline* interview or *Geraldo*'s "Computer Vice" episode, which somehow linked up everything bad and trashy, from serial killer Jeffrey Dahmer having a modem to the 1988 cDc text file "Sex with Satan." Geraldo called cDc "a bunch of sickos." cDc itself touted that and all the other media notices, realizing that journalists would play it safe by calling the same sources that had already been broadcast.

Insiders like *Boing Boing* zine editor Mark Frauenfelder promoted cDc, and the reasonably well-researched 1995 movie *Hackers*, with Angelina Jolie, showed cDc stickers in the background. Some of the time, the media's vague awareness that cDc was about hacking, which was bad, led to bizarre pronouncements. A 1996 story in the *San Antonio Express-News* about the local air force cyberoperations center, for example, hilariously led with the assertion that the unit "defends the nation's secrets from the members of the Legion of Doom and the Cult of the Dead Cow in a battlefield that spans the globe."

Midwesterner Paul Leonard announced an explicit cDc culture-jamming project called cDc Paramedia in August 1996, with the object of "world domination through media saturation." Misha, Kevin, and Luke were enthusiasts of the effort, Luke adopting the title minister of propaganda. Two weeks after the Paramedia announcement, the group wrote: "We intend to dominate and subvert the media wherever possible. Information is a virus. And we intend to infect all of you." Misha cheerfully wrote on the group's site, "We're a neo-Marxist, anarcho-socialist guerrilla unit forged for the sole purpose of getting on TV." The group considered what it was doing to be performance art. Back then, the truth didn't seem as endangered as it does now, so muddying the waters for a cause struck them as ethically acceptable. "It's one thing if you have a state sponsor of disinformation and propaganda that is trying to affect a particular political outcome, versus trying to raise consciousness of some issue that might not break through otherwise," said one member of cDc. "The circumstances matter."

At the time, the group considered getting rid of its old bomb-making recipes out of a sense of social responsibility. But Kevin voted with the majority against burying evidence of the "Anarchy period of the Cyberpunk's progress," as he termed it in a group email. Instead, he suggested adding a disclaimer that would say in part: "If you're smart enough to use a computer and seek out the cDc, then you should be smart enough not to screw around with something like a bomb recipe that is full of spelling and grammatical errors. If the author can't spell or punctuate properly, what the fuck makes you think he can describe how to build a bomb that won't kill you?"

cDc became the first hacker group to issue press releases, and Misha compiled a list of email addresses for hundreds of journalists. Whimsically, Luke took advantage of improper access to various databases and sent printouts to an idiosyncratic list of celebrities as well, including Sean Connery, Harrison Ford, Uma Thurman, and Luke's favorite person, the muscled and campy *A-Team* star known as Mr. T. Meanwhile, the group remained shadowy, using only handles in its communications and public speeches.

cDc's open pursuit of attention struck many hackers as refreshingly candid at a time when other hackers were posing as criminal geniuses or visionaries. They were high-functioning tricksters, the media and their audiences the most common victims. A crowning achievement came after a Japanese television reporter complained that her producers had rejected her thoughtful piece on hackerdom because it lacked excitement. Wearing masks or sunglasses and trying to look scary, Luke and two others agreed to be interviewed on camera telling tall tales. They claimed to be able to divert both moving trains and satellites. "They were the showmen of the industry," Def Con founder Jeff Moss said of cDc. "They were great at taking an issue and calling attention to it." As for truth-telling, Luke saw as his model the Yes Men, politically driven artists who say they use "public spectacle to affect the public debate."

As the Netscape browser and Microsoft's Windows 95 operating system brought the internet to the masses, the security issues that had previously been glaring to hackers suddenly put everyone at risk. The L0pht might flag a few flaws out of thousands that experts could find at any time. But word of even those rarely reached the average computer user. Television commercials funded with venture capital and Microsoft's monopoly profits hyped the amazing online world. But no one had a strong financial incentive to point out the pitfalls. Almost none in the mainstream media covered security full-time, and those who dabbled were under pressure to write about the great advances in computing, which public-relations people also pushed, not the complicated potential problems that their editors couldn't quite grasp. While cDc played with the media's gullibility, it was learning more about how it worked. The group was probing the press the same way it poked at software, and it gradually realized that the greatest threat to security was the poor distribution of true information.

The best place for cDc to start fixing that was at Def Con in Las Vegas. Luke spoke at the third Def Con, in 1995, giving a miniature course in media training. He retold the *Dateline* story, explaining that the correspondent had badgered him over whether he felt remorse and that he had learned a lot from the experience. For the most part, "the media sucks," he warned. "You very rarely see a positive or even accurate view of hackers through the media." Luke advocated sounding out journalists on their angles and declaring what was off-limits. But engaging with serious ones could be worthwhile, he believed, because hackers were in the best position to speak through the media and tell people how to protect themselves and when companies were shipping software full of holes. Public voices were crucial for their kind as well as for consumers, because politicians were weighing laws and enforcement choices that would decide whether hackers would have to stop exploring or face jail.

[x x]

Luke's talk at the 1995 Def Con and other media appearances made him a bit of a celebrity among hackers, which made it easy to meet new people, many of whom wanted to join cDc. But cDc didn't want to be just a social club. That's where cDc's Ninja Strike Force came in. Sam Anthony had dreamed up the idea after taking kung fu classes and was the first leader of the auxiliary group. "Terrible people were interested in joining cDc," Anthony said. cDc wanted to stay small, like the best invite-only bulletin boards. The compromise was keeping cDc elite but expanding through the NSF. So Sam cribbed from a sneaker design, wrote a satiric origin story, and made T-shirts. Early members were people the group liked and respected, including Chris Wysopal,

Window Snyder, pioneer maker Limor Fried, and early Apple and Netscape engineer Tom Dell, who had written software for Mindvox and quietly ran Rotten.com, forerunner of the shock website 4chan.

That year's Def Con had drawn a then-record three hundred people, and at three hundred pounds, Luke was hard to miss. Oakland hacker Josh Buchbinder, who knew him only online, first spotted him in the flesh on the casino floor, holding a teenager upside down by his ankles and shaking him until the coins fell out of his pocket. The kid was so excited that after Luke let him down, he ran away squealing. Someone explained to Josh that it was considered a great honor to be shaken down by Deth Vegetable. That night, Josh joined Luke and his friends to go off into the desert, take drugs, and shoot guns all night, Hunter S. Thompson–style. In a minor miracle, no one was hurt.

Josh stayed in touch with cDc members over the next two years as his skills improved, and in 1997, Dan MacMillan sponsored him for admission to cDc as Sir Dystic. Josh was attending junior college by then after dropping out of high school. He felt behind the curve technologically, since his Bay Area friends were all playing around with Linux, the breakthrough free and open-source operating system that was challenging Microsoft inside big-company server rooms. When Microsoft came out with versions of Windows that could handle internet connections, Josh poked at it. Though his friends thought Windows so inferior as to be uninteresting, Josh figured that enough regular people would end up using it that any research would be worthwhile. What he saw was horrifying. There was essentially no security at all. Anyone who used a Windows machine to read email or browse the web could easily lose control of his or her machine to a stranger. Just about any kind of software would run on the system, and it could be made invisible to the user by those who knew what they were doing. All a user had to do to be infected was click on a file with an innocuous name.

Josh was far from alone in raising the alarm at Microsoft's head-in-the-sand approach. Chris Tucker sent a draft of a rant to the cDc mailing list in 1997, declaring "Microsoft is evil because they sell crap" that only has a chance of getting fixed in a future version if enough people call Microsoft to complain. "You stupid fucks pay Bill Gates to beta test his crappy software," Chris wrote. The problem was compounded because Microsoft sold to a handful of computer makers, not the end users, and Microsoft held all the power in those relationships.

Josh knew he could write a program that would prove the point, that would give invisible control to an email correspondent or anyone else who could establish a connection. He could use such a tool himself, to spy or to steal. But that would break the 1986 hacking law while not being all that much fun. Releasing it into the wild, on the other hand— with as much fanfare as possible—would force Microsoft to admit it had a problem and do something to protect its customers. As it stood, selling Windows 95 and 98 "was like giving loaded guns to children," Josh said. "My point was if we can do this, anybody can. They needed to take this seriously." Plus, with the help from the media, it would be damn funny to watch.

He emailed the cDc list and asked what the other members thought of the idea. Carrie Campbell was opposed to it. She had moved from technical writing to running an internet access provider and now lived near Microsoft's main campus, where she had many friends. Beyond that, she knew that the program would give new power to thousands of relatively unskilled "script kiddies." She saw the public-service argument; she just felt the likely side effects outweighed it. "It's going to hurt average people," Carrie told them. But she was in the minority. The others gave Josh all the encouragement he needed. Just to make sure he wouldn't get slapped in handcuffs simply for writing a malicious program, Josh picked up the phone and called the local FBI office. He asked for an agent in the criminal division. "Would I be in trouble if I released a program that others could use to hack people?" he asked. "You'll have to ask a lawyer that," the agent responded. Josh would not be deterred. "No, you're the FBI," he said. "Would you arrest someone who did that or not?" The agent asked him to hang on. After a while, he picked up the line again. "We would really rather you not do this," he told the hacker, but it's not technically illegal." Josh checked one last time to be sure: "So, I'm good?" he asked. "You're good," the agent sighed.

Then came the hard work: more than a year of prodding for un-documented programming interfaces, the hooks that allowed programs to run on top of Windows. Josh had never written anything remotely that ambitious. But he knew it was possible, he thought Microsoft's security bordered on the criminally incompetent, and he wanted to impress Mudge and his other new friends in cDc. He smoked a prodigious amount of marijuana and kept hammering away through trial and error.

By 1998, Josh was getting a fair amount of encouragement in person. Misha had moved to San Francisco in 1992 and had bragged about it to Luke and the others back east at every opportunity. One of Misha's first contacts was the editor of a magazine called *Mondo 2000*, who reprinted his Information America piece and introduced him to her boyfriend, Eric Hughes, who was about to start the Cypherpunks mailing list, hosted by John Gilmore. Misha spread the word among hackers. The dot-com boom that began with Netscape's initial public offering in 1995 lured more waves of cDc members and friends to California. Dylan Shea took a job at the Mountain View headquarters of Netscape itself in 1996, and when the company offered to pay for his move, he brought out Luke's gear as well. Luke re-created the nonresident part of his Allston hacker group home, New Hack City, in San Francisco, turning cDc into a bicoastal operation. First came a hacker space in an old can factory on the border of Dogpatch, a run-down bit of the city. Then came a spot at Market Street and Sixth Street so rough that Luke once caught a woman hiding behind his bulk to smoke crack on the sidewalk. The label on the apartment directory said Setec Astronomy, a nod to the hacker movie *Sneakers* and an anagram for "too many secrets." At one open house, someone not in on the joke asked why astronomers would be in a basement apartment.

[x x]

cDc believed that Microsoft's response to Josh's program would be directly proportional to the amount of noise it made. So with its greater understanding of the media, cDc wasted no time in building interest in what it had dubbed Back Orifice, a crude pun on Microsoft's Back-Office software. It explained in writing what the program could do well ahead of the actual release, which was planned for the biggest Def Con yet, in 1998. It was up to the hacker how to install the program on a target machine, but it could be combined with any desired executable program, like a word processor or calculator, and emailed to the intended victim. Luke's press release called out features that could log keystrokes on the target computer and encrypt traffic to the hacker who had sent the program. Other software writers could add modules for still more functions. cDc did not advertise the fact that it had taken mercy on Microsoft and the young antivirus industry by setting the default port for inbound traffic as 31337—hacker-speak for *eleet*: that is, "elite." All anyone had to do to stop off-the-shelf installations by non-coders was block traffic to that port.

Luke coordinated major stories with *Wired* and other publications while Kevin and others concentrated on making the Def Con presentation as theatrical as possible. During the Saturday afternoon peak of the three-day con, Kevin and Dylan invented some last-minute gimmicks just before the start of the 4:15 p.m. panel. As Sir Dystic, Josh then went onstage and droned a few boring sentences. A planted heckler, yelling that Back Orifice was a hoax, ran to the stage and grabbed the mic. Luke charged at the man and hoisted him offstage. Then the rest of the cDc crew rushed on. Bringing up the rear was Kevin, with a T-shirt reading GRANDMASTER RATTE, a thick chain around his neck, and white rabbit fur chaps over his jeans. He jumped on the table and started rapping about cDc.

"I can feel the love in the room!" he shouted. "We love our people!" Then he whipped the crowd into a call-and-response: "When I say Dead, you say Cow! Dead!" "Cow!" Kevin handed the mic to Sam Anthony, who sounded calmer. But Sam wore a stocking hat pulled over his face showing a cow skull, and he was explicit about asking the recipients of Back Orifice to hack, and to hack for a cause. "We want you to give back to cDc," he explained. "We are making it so easy that an eight-year-old can make a difference—can fuck shit up." After

Carrie said a few words, Josh took over and ran through the functions, drawing applause when he showed it popping up a Windows dialog box with wording of his choice. He took questions, and at the end the group threw CDs with the program to the crowd. Afterward, Josh did his pre-approved interviews with *Businessweek*, CNN, NPR, and the BBC, all of whom were stunned that he wouldn't give his real name. *USA Today* and dozens of others ran stories in the next day and a half. The *New York Times*, which had already mentioned cDc in a broader Def Con piece, returned with an article on Back Orifice alone, noting in the second paragraph that cDc said it was trying to get Microsoft to focus on security. It also outed Josh as Sir Dystic.

Nothing like this had ever happened before. At the then height of public concern about hacking, at the top conference on the subject, the best-known hacking group had given out a major tool for free. At least in the short term, it certainly seemed like there was going to be much more hacking as a result. "They pulled this joke off on the most dominant commercial force in the world," said Jason Sadofsky. "They wanted to get on TV, and they got on TV."

Yet instead of sounding the alarm or calling for a renewed joint effort to stop hacking or make software safer, Microsoft gave the public impression that it had barely noticed what happened in Las Vegas. "This is not a tool we should take seriously, or our customers should take seriously," Microsoft marketer Edmund Muth told the *New York Times*. The company argued that there were no new vulnerabilities involved in Back Orifice. But that claim was aimed at the uneducated and the media. If Back Orifice had relied on newly discovered holes in Windows or other Microsoft software, the company would have patched them in an update, and the exposure would have been confined to those who did not patch. Instead, the issue was the essential architecture of Windows.

The contrast between what Microsoft was saying and what the more articulate hackers were saying was jarring, and it forced many people to think harder about serious issues for the first time. While "Microsoft is fully buzzword-compliant," Mudge told one interviewer, it only recently had established a security response team and came off to technical people like the town drunk: "It's almost unfair to continually beat up on them, because they can't really defend themselves."

Within months, people had downloaded Back Orifice hundreds of thousands of times just from the cDc website, with an unknown number of pirated versions circulating as well. Many thousands of innocent people fell victim. After internet service provider MindSpring complained that it was detecting at least two new infections of its customers each day, the local Atlanta field office of the FBI opened a criminal hacking conspiracy investigation into cDc and Luke personally, centered on the theory that infected machines were sending stolen data back to servers under cDc control. Because they were doing no such thing, the case sputtered before its formal closure in 2003. Hackers were also among the victims. If they popped a CD with the program into their machines, then read the instructions, they could safely download the virus and begin thinking about sending it to a victim—unless they had set up their machines to automatically play any CD inserted, in which case they immediately infected themselves.

Both fan and hate mail flowed into cDc. One supportive message came from porn star Brandy Alexandre, who said someone had hacked an adult film–industry journalist who used the pen name Luke Ford and deleted files. "Glory be to the dead cow on high!" she emailed, explaining that Ford punished stars by revealing their legal names. "I am your slave if you should happen to repeatedly attack his real names list," she wrote. "What may I do in return, master?" cDc got fan mail from the mothers of teenagers, an NSA staffer, British writer Neil Gaiman, and an actor from *The Texas Chainsaw Massacre 2*. But it wasn't all roses—Josh also got anonymous death threats.

Inside Microsoft, Back Orifice became the company's biggest security headache by far. When the press realized that Back Orifice was big and that the company had no defense, Microsoft came back with a new message: while there was still nothing to worry about, those who were absolutely convinced that they needed the very best security could buy Microsoft's forthcoming, completely reengineered operating system designed for networked machines, Windows NT. That system, Microsoft said, offered "a comprehensive set of security features that make it the best choice for business users' mission-critical applications."

For all of the planning that had gone into Back Orifice, the group was shocked by how big it blew up. A much larger number of people now saw that hacking was a clear and present danger, which was great. But Microsoft's bogus response was still holding the line inside most of its big customers. The businesses had no right to sue over software, since the event when a program changes hands was not classified by the courts as a sale. Well-funded industry lawyers had convinced multiple judges that the electronic terms of service that ended with clicks on "I agree" were for licensing deals. There was no liability for a faulty product, under the law, because there had been no sale; the only remedy was to cancel the license, and that was a dead end. Though Linux was fine for heavy loads, there were few alternatives to Word and Excel for regular office workers.

The more time that went by, the angrier cDc members got. Even Carrie, who hadn't initially supported Back Orifice, agreed it was ridiculous that Microsoft still had its head in the sand after being shown its vulnerabilities. cDc turned to one of its newest and smartest members, Christien Rioux, to take on Windows NT and prove that the group was not a one-hit wonder that had been steamrolled by Microsoft's marketing department. This time, Carrie was all for it.

"They were like the dog that caught a car. They could have stopped," Sadofsky said. "But somehow they got in and drove the car and said, 'Let's see where this goes.'"

> CHAPTER 6

> # ONE MILLION DOLLARS AND A MONSTER TRUCK

A S THE LOPHT and the Cult of the Dead Cow attracted more technologically advanced members, some of the earlier guiding forces faded to the background. Fringe culture fan and cDc cofounder Bill Brown stayed loosely in touch through art college and as he began working on experimental documentary films, some of which landed in major museums. Then he saw cDc become part of the mainstream news. Good for society, he figured, but less of a fit for him. "It is exactly when cDc becomes interesting that I became less interested in it," Bill said. cDc now included the elite of the hacking world, even though its earliest text files had mocked such people as exclusionary showboats. "It became more and more like the thing it was supposed to be pranking about."

Kevin Wheeler sympathized. As the group discussed possible new members in 1999, he lamented: "These guys are all tech guys. Where's the cDc skateboarding team? Why are there no porn stars in cDc? No guys into scary militias and a compound in Montana? Why are we 95% white males?" It was true, cDc was getting less countercultural and less strange. The new tech talent attracted more like themselves—highly educated, curious technologists with a skeptical view of the world. The final crossover member of both the L0pht and cDc was especially that. Christien Rioux's father was a musicology professor in Lewiston, Maine, who brought home programming books as he grappled with software for processing music. Like Mudge, Christien learned to break protections around the childhood computer games in order to keep playing them. The family moved to Monmouth to get Christien into a public-school program for gifted kids. Even so, he skipped eighth grade and spent his senior year of high school at Bates College. Bates had access to Internet Relay Chat and Usenet, and he found cDc text files there in 1992. He was admitted to MIT in 1994, at age sixteen, on a full scholarship.

For someone who had always been the cleverest kid around and had never been to Boston, Christien had a lot to take in. Academically, Christien appreciated that MIT had stopped issuing grades to freshmen after too many suicides. There were parties with other bright kids every Friday, and the newcomer became social chair of his frat. Christien also took responsibility for connecting the frat to the university network, and he closely tracked how the net was developing. He was thinking of himself as a computer game programmer when he read papers by Mudge and others about finding software flaws that could be exploited, and he became entranced with the idea. Among the more promising classes of screwups in programming was a failure to stop what were called buffer overflows. If the coder did not properly limit the amount of data that could be taken into a buffered area of memory, a hacker could enter too much and overflow it, making the excess data delete something in nearby storage. In some cases, that would allow the hacker to take control of the machine. Buffer overflows had been found in a number of high-performance systems, though not the early versions of Windows. Christien found an overflow in Internet Explorer 4, the browser that Microsoft improperly bundled with Windows in 1997 in order to beat pioneer Netscape.

Christien excitedly wrote up his finding for *2600*, which declined to publish it. So he took his printouts to a *2600* meeting at the Prudential Center, hoping to impress the L0pht guys. It worked, and they published an advisory about IE4 under Christien's new handle: Dildog, after the *Dilbert* comic's initial name for the dog character, Dogbert. Microsoft emailed and calmly asked that in the future, the L0pht hold off publishing details of security flaws until a patch was ready. "Maybe that's not a bad idea," Chris Wysopal said. Before that, if companies had complained about being taken by surprise, the L0pht had given a canned response about caring for the users, not the vendors, Christien said. But he found it hard to argue that most users wouldn't be in better shape when a patch was out. So the group began negotiating with Microsoft and other companies. It would offer a month's notice before going public, while the companies asked for more time. Often they reached a compromise in the middle, and the current standard of coordinated disclosure began. Reading the disclosures made it easier for malicious hackers to learn most of what they needed to launch an attack based on the flaws, but everyone who patched right away would be safe. Without the disclosures, only the hackers who took the effort to reverse engineer the patches would have been able to launch the attacks, but there would have been less public awareness of the problems. Mudge and Wysopal, who wrote many of the advisories, became the most visible and articulate explainers of the researchers' side. "I wanted the L0pht to be *Consumer Reports* and Rachel Carson and Ralph Nader," Mudge said. "That was my vision."

Despite his youth, the group took Christien along for hangout sessions at New Hack City, home to the cDc servers. Mudge impressed him while playing quarters by rolling the coins off his nose before they bounced into a beer glass. As an MIT junior, Christien took a class on social issues in computing that turned out to be mostly about security. An early assignment was to look into buffer overflows, and the instructor put up a slide of one attributed to Dildog. "This is going to be much easier than I thought," Christien said to himself. The L0pht invited him to join in late 1998, after Christien graduated, and it used money from the sale of its security tools to pay him to write the next version of its best-known program, the L0phtCrack password breaker. It was a major improvement, brought in almost $500,000, and prompted the squad to make Christien their first full-time employee. "The point of L0phtCrack was to get everyone out of their day jobs," he said.

By the time Wysopal brought Christien in, the L0pht was already famous. *Wired* and the *Washington Post* had written about it, as the advisories and tools drew attention to the downside of marketing-driven technology companies with no legal liability and little market punishment for the insecurity of their products. No one else had enough of an incentive to point out the emperor's lack of clothes.

[x x]

Inside Microsoft's biggest customer, the federal government, Richard Clarke was getting nervous. Nobody seemed to be talking about the risks of hacking. A counterterrorism expert on the first President Bush's National Security Council, Clarke was named the NSC's national coordinator for security, infrastructure protection, and counterterrorism by President Bill Clinton in 1998.

Everything important in the country ran on software, most of it procured on the open market, and yet Clarke kept reading about hackers having their way with it. Surely rival governments could be doing that in America as well. Clarke's suspicions intensified after the war game exercise Eligible Receiver was run by the Defense Department in 1997. An NSA red team, tasked with breaking into Pentagon networks, ran roughshod over them using only conventional tools. Clarke didn't know it at the time, but Moscow was already doing the same thing for real in an operation later discovered and dubbed Moonlight Maze. The success of Eligible Receiver prompted the Defense Department to set up the Joint Task Force–Computer Network Defense, which would work on behalf of all branches of the military.

Yet the leaders of the NSA were still telling Clarke there wasn't much to worry about. He met with the CEOs of Microsoft, networking king Cisco, and database giant Oracle, and they said the same thing. "They were all telling me their shit didn't stink, and I had a hard time reconciling the fact that Oracle and Microsoft and Cisco were all perfect with the fact that all these hacks were occurring," Clarke said. "It seemed obvious I needed to talk to the people doing the hacking. But they [were] probably criminals, so I asked, are there people like that who are not criminals?" Clarke talked to an FBI official who had transferred from the Boston office. "He called back a few days later, said the

Boston office knows this group of hackers that they vetted, they think they're clean, and when they have technical questions they ask them stuff." Clarke took a crew from the NSC up in early 1998. The L0pht gang suggested a drink at a bar called John Harvard's, then watched silently to see how many officials were there and how long they would sit before getting up to leave. After an hour, when they finally stood, Mudge said hello.

After beers, the group invited the NSC team back to the L0pht. The men showed off a little of what they were working on. As they were leaving, Clarke huddled with the other officials in the parking lot. A bit spooked, the hackers told Mudge to tell them it was impolite to whisper in their presence. He marched over and did so, demanding to know what the group was discussing. Everyone looked at Clarke, who looked at Mudge frankly. "We were saying we thought all this wouldn't be possible without some government's support," he told the longhair. "Have you gotten any?" No, Mudge replied, then joked: "If you have an offer, we'll listen." Clarke paused, then laughed.

Clarke stayed in touch with Mudge afterward. Out of the group, Mudge was especially receptive. Some old-timers in cDc still instinctively disliked the government, or at least some of its laws, such as the ham-handed Computer Fraud and Abuse Act, or certain branches, namely the FBI. But Mudge was among those whose families had been paid by government dollars, and he had worked as a government contractor at BBN. He also felt that everyone should know what he knew. Perhaps the government would still make the wrong calls, he realized. But at least it wouldn't be out of ignorance. There was one last, less noble reason for playing ball. He was counting on people in the military being able to vouch for him if the FBI got overexcited and suddenly raided the L0pht. "Should I find myself in court with a bunch of L0pht folks, brought up on charges of disturbing the peace or something, I wanted to be able to reach out and have a bunch of people with uniforms and a bunch of medals sitting up there as character witnesses," Mudge said.

Clarke was quietly preparing an order that would be known as Presidential Decision Directive 63 on critical infrastructure protection, which gave the government more authority to lead on private-sector

security measures. For ammunition in the intergovernmental turf battles and to head off complaining from the US Chamber of Commerce,
Clarke called on his new ally in Boston, and shortly after, Senator Fred
Thompson formally asked the seven current members of the L0pht
to testify before Thompson's committee about threats from hacking.
Mudge said they would only do it if they could use their hacker names,
which was all that had been public about them up to that point, in order to protect their day jobs. Thompson agreed. The National Security
Council had a message they wanted to get out, and this was an opportunity for the L0pht to interact with the government "without us being
labeled as criminals," Mudge said.

Everyone who didn't already own a suit bought or borrowed one,
and they testified in May 1998. Chris Wysopal, Brian Hassick, Joe
Grand, and three other members of the L0pht sat with Mudge and his
heavy-metal hair in the middle. He was the only one present who was
in the Cult of the Dead Cow. Dan MacMillan had moved west, John
Lester had been displaced from the L0pht, and future cDc member
Christien Rioux wouldn't join the L0pht for a few more months. "If
you're looking for computer security, then the internet is not the place
to be," Mudge told the senators. "How can we be expected to protect
the system and the network when all of the seven individuals seated
before you can tear down the foundation that the network was built
upon?" The most dramatic claim they made during the testimony was
that they could take down the internet in thirty minutes through a problem they had found in the internet's routing procedure, Border Gateway
Protocol. The L0pht had already contacted the relevant manufacturers
about the issue, Mudge said later. The senators present were far more
alarmed by what they heard from the hackers than they had been by
what the military and intelligence brass were saying. "We were a visceral representation of what the adversarial view was," Wysopal said.

[x x]

The testimony made the L0pht into the first group of rock-star hackers, and Mudge was the bandleader. But even with cover from the

government, he and others in the L0pht, especially the newer and cleaner members like Wysopal and Christien, were nervous about their plans to not just improve the state of security but earn a living in the process. They knew the L0pht couldn't make things up or throw verbal hand grenades at the government or giant companies.

For that, there was the bad cop, cDc, which played with seeming sinister. Both groups were mad that Microsoft had sidestepped Back Orifice without getting serious about security. It sent two messages: that Back Orifice was not a problem, and that to the extent some users felt it was, they could always just switch to Windows NT or later versions. The only way to fight back was to create a new version of Back Orifice that could beat the new operating system. That would show that Microsoft's main programs remained fundamentally broken because they did not give users a reliable way to know what was trustworthy on their machines. Christien was the best person to write the 1999 sequel to Back Orifice, and he had been drafted to cDc earlier that year. Though he was being paid by the L0pht at the time, the L0pht could not publish what they decided to call Back Orifice 2000, because that would have tied the L0pht too closely to cDc and therefore to Def Con and costumes and rapping, along with the air of drugs and crime. "BO2k couldn't have been L0pht, because it already had stank on it," Mudge said. It had to remain separate from the L0pht to avoid alienating the Richard Clarkes of the world, who were potential L0pht customers and partners. cDc would release BO2k with even more spectacle at Def Con in July 1999 than it had created the year before.

Though Windows NT had been put together much more carefully than Windows 98, the core problem remained. The computer handed off too much control to outside programs that were not signed digitally, or otherwise attested to as authentic, by Microsoft or the vendors of those programs. As a result, it was only a modest challenge for a hacker to get a malicious program running on a Windows machine, then hide the fact that it was there. cDc wanted to warn everyone that Microsoft's security setup was too complex and that users could be running something dangerous without knowing it. It wanted the company to require customers to verify the source and condition of the outside software, so they could then decide what to trust. "Our position should not be one of Microsoft bashing, but rather of user education," Christien wrote to the others in the group. Microsoft had screwed up, and it "needs to take responsibility for putting so much power in the hands of the user."

Christien's program was dramatically better than Josh Buchbinder's. Beyond the coding improvements, cDc wanted to settle an argument by Microsoft supporters and some hackers that Back Orifice was unsafe and might contain a back door for cDc, neither of which was true. This time, the group wanted to release the code, to make it open-source. That would prove that there was nothing up cDc's sleeve. It would also up the stakes by making it easy for hackers to modify it, rendering antivirus programs, which look for identical versions of things previously flagged as bad, far less effective. The Atlanta FBI office warned the Pentagon and other potential targets that the new version would be "potentially more destructive and difficult to eradicate" and that all concerned should "aggressively review and monitor" their security measures. The Defense Department's Criminal Investigative Service looked into BO2k to help the military develop countermeasures but did not pursue the matter beyond that. There would be more hacking. But that would put more pressure on Microsoft to make deeper fixes to its programs.

As before, cDc drew the line at integrating any additional program that would take advantage of a software flaw to deliver and install the tool. The lack of such an exploit limited the group's moral culpability, cDc felt. The members were distributing a safecracking tool but not the keys to the vault that held the safe. There was also a risk of legal liability. Courts had by then ruled that code was speech, and therefore almost no regulation could stop it from being written and distributed. But most serious programs also used encryption for communication. In the case of BO2k, the encryption would prevent data that was moving from an infected computer to the hacker's computer from being intercepted and deciphered. The Clinton administration had continued to clamp down on the export of strong cryptography, over the objection of US-based multinational technology firms. The government likened serious encryption to a weapon, albeit a defensive one, and made it subject to export controls. As is still the case, Washington wanted to maintain its ability to break codes used elsewhere. If strong encryption products go to other countries, that makes it harder. So Uncle Sam has used a variety of regulations to stop or hinder such exports.

Christien did not want to get in trouble with the US government. A lawyer was hired to take a look at the program and make sure they wouldn't get in export trouble. She advised them to put a little more effort into keeping the program out of enemy hands, at least until some pending court cases over similar issues were resolved. She told them to check whether the downloaders' Internet Protocol addresses were in the US and to have the locals promise not to transfer the program beyond the border. Those outside the US got a version with less encryption. "The first rule of the activist is to not get caught," Kevin wrote to the list. "Federal time is bad time."

In a new experiment, cDc decided to call BO2k a tool for remote administration in its press kit and documentation. In effect, it would be making the argument that the software was the most sophisticated tool for surreptitious electronic break-ins while also being among the best tools ever made for corporate staffers to remotely monitor what was running on office computers and install new programs. While Symantec and Compaq charged more than $100 for remote tools, cDc would offer similar or better capabilities for free, with code the user could examine. If the group pulled it off, it would be dropping the pants of not just Microsoft but also the established security companies, which cDc felt were getting fat on the internet stock-market boom while peddling average products.

The security establishment showed its true colors as launch date neared. Atlanta-based Internet Security Systems, which had first sold stock to the public in 1998, railed against the dangers of the impending BO2k to drum up its own business. But behind the scenes, it was sweet-talking cDc and asking for an advance copy of the program. That way, it could claim to be blocking BO2k before it came out. An ISS intermediary even offered cash, which was a terrible way to approach a group of volunteers who were convinced they had found the moral high ground. "ISS is just flat-out sleazy in a lot of ways," Mudge said then.

cDc exposed the offer and sent a reply it leaked to the press, saying, "We are gladly willing to provide you with the software you desire if and only if you will, on exchange, grant us one million dollars and a monster truck." Some ISS employees worked in the office through the Def Con weekend, and they sent kids to the show floor to grab a CD and upload it at the first possible moment.

Because cDc wanted maximum impact, it needed maximum press. For that to happen, it had to have a touch of evil, Kevin reasoned, the same way a punk or metal band craved condemnation. "The hip press has to love us and the square press has to hate us for this to work. That's the eternal conflict in society to play off and ride," he wrote to the group. "The day [evangelist] Pat Robertson says something positive about cDc is the day we're over. The conflict, the drama is what makes this interesting and worth writing about."

The insider appeal helped as well. When the group made jokes only other hackers got, it gave it street credibility and simultaneously impressed the outsiders, who realized cDc resonated with real hackers more than the people in suits did. But Kevin warned the group not to get cocky, reminding them that cDc had started by mocking the Legion of Doom and other self-serious coders. The point was to have fun and be useful. He wrote just before Def Con, "If we fall for our own hype, that's the same pathetic retard bullshit that useless rock stars and movie stars fall for, when they 'can't handle the pressure' and get some debilitating drug habit or become assholes, 'cause they don't understand their role in a system."

[x x]

When the big day came, the presentation began with electronic music and the recorded voices of a farmer ordering his daughter to put a cow back in the barn and the girl refusing. With lights pulsating on the screen in the darkened room, Kevin rapped rhythmically and paced the stage for more than five minutes. It was the first full rock-and-roll hacker release, complete with stencil spotlights beaming the group's longhorn symbol. "cDc loves you!" Kevin shouted, and once again led a call-and-response chant: "Dead!" "Cow!" "Kicks!" "Ass!" Even after the house lights came up, he went on, performing a mock faith-healing and calling for amens from the thousands in the audience. Finally exhausted, he asked Sam "Tweety Fish" Anthony to introduce the rest of the nineteen cDc members assembled onstage, by far the most in any one place.

"This is Deth Veggie, you all know him," Sam began. "The future of programming, Mr. Dildog." He went through them all, ending with the surprise reappearance of cofounder Bill Brown, wearing an old-fashioned suit. Then Sam said he needed to make two amendments to his year-old exhortation to go out and hack. "Pick a cause" instead of hacking at random. And don't get caught. Christien then gave a straightforward demo of the core product and some of the available additions, stressing that the code was thoroughly customizable. The crowd repeatedly interrupted him with applause and oohs and aahs after he explained features, like the ability to delve into other machines connected to the target. After he and Josh fielded questions, the lights suddenly went out again. Bill pulled off his tear-away suit from the front, revealing pasties on his chest. Mudge played rapid licks on the guitar and smashed it against an old PC.

Christien had burned advance copies of BO2k onto CDs with a machine owned by Ninja Strike Force member Limor Fried, whom he was dating. Unfortunately, her machine had been infected with a virus known as Chernobyl, which spread to the CDs for the press and those the group brought to Las Vegas to toss to the crowd, which included diving ISS employees. Once hackers at Def Con uploaded it to the net, someone detected the virus and cried foul. cDc once again faced suspicions of hacking fellow hackers. cDc admitted the screwup and apologized. Fortunately, the version available for download from the cDc website was clean all along.

Christien was so young, and had come to cDc so quickly, that he didn't have connections to criminals like some of the others. When people asked him whether malicious hackers would use his creation for crime, Christien said he didn't think so. In retrospect, that was implausibly naive. Though far from malicious himself, he said he wasn't at the pious extreme, either, but a "question-asker. I am not completely white hat, because I wasn't trying to secure the world, but to raise awareness."

Kevin's hometown paper saw nothing but black hats, and Swamp Rat couldn't have been happier. "We prefer to call it what it is— organized crime and terrorism," the *Lubbock Avalanche-Journal* declared in an editorial condemning BO2k. "BO2k is a weapon. It has no useful purpose other than to attack and destroy the property of an individual or corporation. We believe that it is time for an aggressive campaign against organized hacking. We find it disgraceful that a weapon like BO2k can be given an in-your-face public release by CDC without any real fear by the group or its members of being held accountable." As Kevin recapped it for friends, the paper was "practically calling us godless commies and a threat to the American way of life and their daughters' virginity. It was fuckin' beautiful."

Security companies didn't go as far as all that, but they generally categorized BO2k as a virus. Finland-based F-Secure noted that it was likely to be used by hackers, especially since the program went to such lengths to run without detection and to avoid being deleted. It kept changing its process identifier and created new processes in case one was killed. The best-known cryptography expert of the era, Bruce Schneier, gave it a qualified thumbs-up. He wrote that it was useful for systems administrators. He also acknowledged that miscreants would love it, since BO2k was "one of the coolest hacking tools ever developed." Schneier openly addressed the philosophical game that cDc was playing, and he declared it a winner. "Since it is not distributed by a respectable company, it cannot be trusted. Since it was written by hackers, it is evil. Since its malicious uses are talked about more, its benevolent uses are ignored. That's wrong," Schneier wrote on his blog. He said that Microsoft security was virtually nonexistent in Windows 95 and 98 and that a user would have to make more than three hundred adjustments from the defaults in Windows NT to make it safe.

Microsoft had created the danger, and "what Back Orifice has done is made mainstream computer users aware of the danger. Maybe the world would have been safer had they not demonstrated the danger so graphically, but I am not sure," Schneier wrote. "Microsoft only responds to security threats if they are demonstrated. Explain the threat in an academic paper and Microsoft denies it; release a hacking tool like Back Orifice, and suddenly they take the vulnerability seriously." Some of the most enthusiastic support came from those high up in government and defense contractors. One Lockheed Martin expert wrote to a security mailing list that the fanfare around Back Orifice had prompted him to look into the prevalence of Trojan programs, which allow computer takeovers, and that he had been stunned to find more than ten in quiet circulation. He said the new noise around BO2k was the shock treatment that network administrators needed. "If your security is not strong enough to stop script kiddies with publicly available tools, then you have no hope of securing your network from professionals waging war," he wrote. "Wake up people, it's going to get much, much worse."

In public, Microsoft again pooh-poohed the issues, even after hackers posted videos of themselves taking over strangers' machines. In private, it panicked once more. An executive asked security worker Rob "Whitey" Beck, a friend of Carrie Campbell, to have her bring in a video of the Def Con presentation. Carrie wanted to help Microsoft do better. So she walked across the street to campus, met the executive, and then gasped as he took the CD she handed him and popped it into his office computer. "Wait," she said, before he could type "run." "Do you have a sandboxed machine?" She meant one where a malicious program couldn't move to other computers. The man stared at her. "You're not putting that CD into a machine connected to the network, are you?" Sure, he said. "Um, really? Don't you have a separate box you can put that into?" Another blank look.

"So let me get this right. You have a member of an internationally famous hacker group that just released a tool to help people defeat Microsoft security, sitting in front of you, you don't know her at all, and you're putting a homemade CD she handed you directly into your machine? Please tell me you at least are running an antivirus tool against it?" He wasn't.

All the noise still hadn't fully penetrated Microsoft. But it had finally gotten through to Microsoft customers, especially banks, who pressed the company to make serious changes or risk losing them all to Linux. After BO2k, Microsoft did more to promote the use of digital signatures that established who was standing behind a program. "File integrity became a big thing" too, Beck said, with software that checked that a program had not been altered. Security budgets rose across the industry as companies spent more on deeper security research and bought firewalls and intrusion-detection systems.

Pulling off feats like the Orifice launches two years in a row cemented cDc's position in security culture as the internet boom was peaking. In a format later adopted by Reddit for its AMAs, the leading tech-discussion site Slashdot arranged that fall for cDc to answer reader questions under their various handles. Amid a lot of joking and posed crudeness, they articulated quite a few beliefs and goals for security that had many tech-industry readers nodding. They especially wanted software companies to put more thought, effort, and money into user safety and privacy, even if they did not consider themselves to be in the security business. "Make security concerns and security audits an integral part of the development process, rather than an afterthought," urged Sam. Added Christien: "Encrypt everything. Eliminate HTTP and go right to HTTPS everywhere." About nineteen years later, Google's Chrome browser would finally begin warning users who reached HTTP sites that they were "not secure."

Since they all had day jobs, they laughed off suggestions that they distribute a greatly expanded suite of software, but they actually had more ambition than they let on. They had already begun following a path proposed by the member urging them to use their notoriety for the greatest possible good: Oxblood Ruffin.

> # CHAPTER 7

> # OXBLOOD

LAIRD BROWN WAS the truest outsider to be welcomed into the Cult of the Dead Cow, and yet he would have the greatest impact on its trajectory. He was a Canadian-born internationalist in a group dominated by Americans, and a modest technologist who joined at a time when cDc was attracting some of the best minds in security. Laird brought two things: a more refined style of Kevin Wheeler—quality marketing and a sense of moral urgency. Kevin and Bill Brown had always held that cDc was not about technology itself; it was about connection and communication. As it probed technological issues more deeply, the group became more frustrated with the way companies and government officials were acting. The big companies ignored problems unless they were exposed so badly that customers threatened to leave, which was rare for a monopoly like Microsoft. The security industry was not fixing things because the fundamental problems ran deeper than software: it was about business models, corporate power, and legal limitations. And the government was oblivious, slow-moving, or bought off, especially outside of the military. cDc had absorbed all of that, and with its newly bestowed rock-star status, it was ready to take the argument to a more radical place.

Tall and garrulous but cerebral, Laird had shown up before the Back Orifice releases. He had read John Lester's personal account of hijinks at HoHoCon '94, and he emulated its style. He spoke cDc's own language and gradually convinced its members to stake out a broader position. Because he understood where the group had been and where it was heading, he had the answer to their nagging sense of frustration. He began with a lighthearted, oddball, flattering email out of the blue to Luke Benfey at his L0pht address in September 1995. "Cher legume," it addressed Veggie, "I know your travails . . . the many calls upon your fertile resources. . . . Alas, it is part of the burden of greatness. That is why it pains me all the more to elicit your teaching." Using Latin as well as French and his own version of self-mocking leet-speak, Laird said that he had spent two days reading cDc's archived text files and was terribly sorry to bother him but wanted to know if there was anything else nearly as hip that Luke could recommend. The emails came sporadically over the next year, usually to one or a few members of cDc, who forwarded them to the whole list. Laird said he was working for a not-for-profit technical consulting group with a lot of Canadian government contracts.

Laird came by his sense of ethics, disdain for authority, and showmanship well before college. Born in 1950 to a welder father and teacher mother in the Toronto suburb of Hamilton, Laird was a nominal Protestant. But he attended a boys' Catholic high school and enjoyed the clear moral framework of those around him, including strong support for the civil rights movement in the US and for Vietnam protestors, many of whom fled to Canada to avoid the draft. "It was a defining moment. All these things appeared to me to be moral evils, especially [denial of] civil rights," Laird said.

He played classical violin from early childhood and performed on a range of instruments in various genres for money while studying music at the University of Windsor, until the studying took all the fun out of it. After jobs as an auto factory worker, cook, and photographer, Laird moved to New York. There he edited insider newsletters at the United Nations and then, assisting a State Department retiree who had intelligence connections, compiled a multivolume compendium about the inner workings of the UN. "I read a million documents and found out who everyone was," Laird said, developing deep knowledge of the ideals and practicalities there. Then he consulted for West African and South American countries, explaining how things worked at the UN. He stayed until the Libyan mission offered him a lucrative job as spokesman. It would have been ridiculous to accept, but the offer prompted self-examination that ended with Laird leaving the city and moving back to Toronto. During his decade at the UN, the predicament of Chinese dissidents haunted Laird. Market liberalization in the 1980s had helped spawn a student movement in China for greater freedom of speech and democracy, and the Communist Party wavered on how to respond. After as many as a million protestors gathered in Tiananmen Square in 1989, Premier Li Peng declared martial law and sent in troops, who killed more than one hundred. Liberals were purged from Communist Party leadership, and the range of permitted discussion topics narrowed sharply. Still, Laird's activism began only when he joined cDc in 1996.

Befitting someone who had labored for years among the silver tongues at the UN, Laird's tone remained respectful even as he became a part of the group. But he gradually began cajoling cDc for change, with one underlying point and a well-chosen target. The point was that cDc was famous but did not stand for much that was vital beyond tech security. And the area to expand the vision, he suggested, involved the Chinese government. It was a deeply personal argument because Laird had traveled in Asia and was close to people fighting for human rights in China. He also said that during his time at the UN he had met Chinese diplomats who had hinted at unhappiness with events in their homeland. From the early days, Laird told the others about a guy he had met back in Toronto, a Chinese exile helping others get out in the wake of the massacre in and around Tiananmen Square. Gradually the story got more elaborate. The friend was protected by mobsters who smuggled people for other reasons. He had a network of helpers. And he was interested in using technology to help dissidents, which was conveniently right up the alley of cDc.

The house rules of cDc said that everyone would get a chance to weigh in on a candidate for membership but that Kevin would have the final say. In addition, someone needed to meet him in person. Luke visited Laird in Toronto in the summer of 1996, and Laird was admitted to cDc not long afterward. For his handle, he picked Oxblood Ruffin, combining a reference to the oxblood-colored Doc Martens boots popular in the British punk scene with a nod to David Ruffin, lead singer for the Temptations on songs including "My Girl" and "Ain't Too Proud to Beg."

In October, Luke returned to Toronto with John Lester and Sam Anthony in tow. Laird memorialized the event in classic cDc style, with a funny, text file–style email to the rest of cDc announcing that they had conducted the First Annual Won Ton Con at a Chinatown restaurant. He described the restaurant as a favored hangout of the Hong Kong Blondes, "a pairing of Chinese computer scientists and democratic activists" who he said could not join them that day for security reasons. A few months later, Laird gave an odd internal backstory for the Hong Kong Blondes. He emailed the others that he had invented the group as a joke, but his unnamed boss at the not-for-profit web consultancy had been "fascinated" by the fiction, "this great mythical force on the net" that could spawn imitators and confuse the Chinese government. Laird told cDc that he'd introduced his supervisor to the exiled dissidents in Toronto and that the Blondes had become a reality.

The Chinese government provided the perfect catalyst to push cDc into politics. It hated the free flow of information, a core value of cDc and the hacker movement it helped lead. China also naturally opposed the US government, where some of cDc and many of their friends and relatives worked. And China was doing business with the same companies cDc loved to hate, chief among them Microsoft.

Laird was a master marketer, and his cause sharpened his drive. Though his mysterious arrival and vague background perplexed the group, "Laird spoke human," Misha Kubecka said, and that was a big step forward for the geeks. No matter what, he would find a way to tell a compelling story that would hook the media, security practitioners, and perhaps mainstream technologists. "Thank goodness we had Laird's guidance," Carrie Campbell said. "He said, 'You have a little window

of fame right now, what do you want to do with it? Do you want to run
around like idiots or get something done?'" Laird was becoming the
new wise elder, the role Chris Tucker had played.

[x x]

Like Chris Tucker, Laird wasn't coming from nowhere. He was building
on the politicization that had been expressed most dramatically earlier
in 1996 by the Electronic Frontier Foundation's John Perry Barlow, a
libertarian Republican. While a party had raged on around him during
the World Economic Forum in Davos, Switzerland, Barlow had read
that an over-the-top attempt to ban web porn had just been signed into
law in America as part of telecom legislation.

"A Declaration of the Independence of Cyberspace" was Barlow's
over-the-top response. A deliberate echo of Thomas Jefferson, it began with a hint of Karl Marx: "Governments of the Industrial World,
you weary giants of flesh and steel, I come from Cyberspace, the new
home of Mind. On behalf of the future, I ask you of the past to leave us
alone. You are not welcome among us. You have no sovereignty where
we gather." The sixteen-paragraph war whoop would soon be posted on
tens of thousands of websites. "We are creating a world that all may
enter without privilege or prejudice accorded by race, economic power,
military force, or station of birth. We are creating a world where anyone, anywhere may express his or her beliefs, no matter how singular,
without fear of being coerced into silence or conformity," they said.

"In China, Germany, France, Russia, Singapore, Italy and the
United States, you are trying to ward off the virus of liberty by erecting
guard posts at the frontiers of Cyberspace. These may keep out the con-
tagion for a small time, but they will not work in a world that will soon
be blanketed in bit-bearing media. We will create a civilization of the
Mind in Cyberspace. May it be more humane and fair than the world
your governments have made before." It was idealistic, more than a bit
silly, and remarkably naive for a technology culture that was already
rewarding exploitation of click-happy human behavior.

Twenty years later, Barlow said that the innocence was a deliberate pose. "I knew the arrival of the net was liable to be as powerful in a very negative way as it was powerful in a very positive way. If it was possible for everything to be known for everyone curious about it, it was also going to be possible for just about anyone everywhere to devise turnkey totalitarianism, where they could flip a switch and see everything you are up to." Barlow wanted to "set cultural expectations," he said, to strengthen the side of righteousness for the battles to come. "I wanted people to think and feel that what we were entering into was a golden era, and that it was about freedom, and that it was about the explosion and dissemination of knowledge. And with any luck, we would figure out how to deal with the horrible part as those situations arose."

For all its calculated omissions and excesses of passion, Barlow's howl resonated with a burgeoning crowd of technologists, aspirants, and consumers who badly wanted the government to do anything other than screw up the greatest invention of their lifetimes. The biggest fans, as a class, were the programmers, the people creating technology daily for themselves and for others. By definition, they were all always at work on something unreleased that was going to be better than what had gone before. Within that group, the greatest enthusiasts were the hackers, the nonconformists and explorers who took things apart and put them back together in different ways, and who were the most likely in the world to, in the process, break laws like the Telecommunications Act, the Computer Fraud and Abuse Act, and the Digital Millennium Copyright Act.

Given their tendencies to work in isolation and reject social norms, it is hard to generalize about hacker beliefs. A great many tinkerers did their best to ignore large chunks of the outside world, especially the parts devoted to politics, and some did not pay much attention even to hackers working in adjacent spaces, like other hardware or operating systems or applications. But it is fair to say that most of those who were paying attention to the political world—a number that would grow dramatically as Barlow's half-imaginary independent cyberspace clashed more with the reality of government—were on his side.

As Barlow's declaration reverberated inside cDc alongside Laird's railing about China, Misha would invent the term *hacktivism*, a portmanteau of *hacking* and *activism* and a concept that would play an enormous role for decades as hackers explored their role in society. "The word describes what the Hong Kong Blondes and cDc are doing together: Hacktivism," Misha wrote to the group.

[x x]

The following year featured an August revival of a hacking conference put on in New York three years earlier by *2600* magazine, called Hackers on Planet Earth, or HOPE. The name, updated for the 1997 edition as Beyond HOPE, was emblematic of a conference that would focus more on idealism than Def Con did. It attracted two thousand people, double the 1994 crowd. cDc members had multiple speaking slots, including a Saturday afternoon L0pht panel featuring Mudge and a Sunday afternoon panel with eight from cDc. It was the first hacking conference of any kind that Laird had attended, and by this point he had managed to latch on to cDc and take a surprisingly solid lead role within the group.

Luke played emcee for the half-hour panel, introducing by their handles Chris Tucker, "über-hacker laureate" Mudge, Sam Anthony, "foreign minister" Laird, Carrie, and John Lester. Though the panel was conducted as a general update on the group's activities, Luke made clear that the most important development was a "strategic alliance" it was announcing with a Chinese prodemocracy group called the Hong Kong Blondes, which included technologists and activists. He then turned things over to Laird for his public debut as Oxblood. Clean-cut and short-haired, Laird was the only one on the panel who seemed like he would have been at ease in a suit, though he instead wore a yellow sports shirt.

Laird said that when he was working at the UN, he had met Chinese dissidents who were abroad as far back as 1989 and had stayed in touch with them even after that year's slaughter at Tiananmen Square drove the democracy movement out of view. He explained that the Blondes' name referred to slang for gold, which was seen as a cornerstone of freedom, and described some of what life was like under the repressive regime. He seemed to take credit for giving the Blondes the idea of using the net to help coordinate protests: "A couple years ago I was asking one of my contacts from Princeton, 'Do you guys use the net for any of your advocacy?' It sort of struck a bell with him, and he said maybe that's not a bad idea. The next thing we know, he was getting in touch with some of his colleagues and his associates who were computer scientists and also very sympathetic to the democratic struggle," Laird said. The Hong Kong Blondes had formally started in September 1996, with cDc offering advice. "The hacking community is an international community," he said in his six-minute talk. "We're all in the same community."

Chris "Nightstalker" Tucker urged hackers in the US to be more active politically and educate lawmakers, and Sam said that those well versed in tech could have a major role in at least calling attention to the plight of the Chinese and others: "We have greater power than anyone else, those of us who understand this technology, to disseminate the information. And when the information is out there, it helps." The group then fielded questions about the Blondes, security, text files, and an open-source operating system project led by an associate. For conference chief and *2600* founder Eric Corley, cDc had the perfect upbeat activism he wanted. "They had fun and conveyed an important message," Corley said. "cDc was unique."

The New York appearance gave the people who had traveled a chance to catch up with the local cDc members as well as each other. Laird had met very few of his fellow members before that weekend; they were just electronic pen pals and collaborators. Chris Tucker had never met Sam or Carrie. An especially joyful reunion occurred between Carrie and another old-timer, Psychedelic Warlord. Warlord had gone east for college and toured in a punk band during two of his college summers. Carrie had housed him and his bandmates and fed them when they reached Seattle.

Now in New York and working at an internet access provider, Warlord was out of hacking and didn't attend the formal events. Instead he met the group at the conference's home at the Puck Building on Friday night and came out with them for a party. Carrie, sporting short blond hair and black lipstick, introduced Warlord to John, Sam, friend of the herd Limor Fried, and others who had joined cDc after his time. They talked about the old days, and Warlord wondered if cDc home Demon Roach Underground was still working. One of the others dialed Kevin's modem number and held up the phone to let Warlord hear it try to connect, proof that it was still humming along.

[x x]

Oxblood had the spotlight for just a few minutes of a group panel on the last day of the conference. Most of those in the audience were interested in hacking, not Chinese politics. The mainstream press had not yet turned sustained attention to security, and cDc had not yet won the fame that would come from Back Orifice. So there was little media attention. One young reporter, Arik Hesseldahl, was intrigued and kept after Oxblood. A half year later, he wrote a short piece in *Wired* magazine about the Blondes, passing along Oxbood's new claim to him that they had disabled a Chinese satellite. "Given the wild history of how hackers had so readily cross-pollinated with antiestablishment and counterculture types in the West in the 1970s, it wasn't much of a leap to believe, though I think 'hope' is a more accurate word, that similar things were taking place in China," Hesseldahl said later.

While most of the active cDc members were excited about Josh Buchbinder's work on Back Orifice, Laird kept talking to Hesseldahl, who pressed him for an introduction to any member of the dissident group. Laird demurred but said he could ask questions on the reporter's behalf. That developed into a full-fledged text file in the form of an interview between Laird, writing as Oxblood, and the dissident, whom he dubbed Blondie Wong.

Laird wrote that the conversation had taken place at a Toronto dive rock bar called Ted's Collision and Body Repair. Their talk centered on Blondie's underground network of technology-savvy rebels in China, which had grown by twenty members in the previous year, thanks in large part to advice from hacking groups, including the Cult of the Dead Cow. "When I understood how far the Cult of the Dead Cow reached into the hacker world, and how things were organized, I was able to take the best and use it for our struggle," Blondie said. They chatted about *Seinfeld*, Bruce Lee, fashion, and high-school alienation. But the familiar bantering sucked in casual readers in order to dump them in dark territory: the murder of Blondie's father by Mao's Red Guards and the massacre of student protestors at Tiananmen. Blondie said the brutal repression convinced him to stay abroad and work to protect his compatriots at home. He asked readers to educate themselves, to keep trade relations contingent on improvements in human rights, and to expose or even hack American companies doing business with China. "If people want to participate, they should use the skills that they have," Blondie told Oxblood.

Laird gave an advance version of the interview transcript to Hesseldahl as exclusive material, knowing that would make the article more appealing. Hesseldahl pitched his follow-up story to *Wired* again, but it demurred, and he ended up with a deal at online spin-off *Wired News*. The story traveled from there far more than it would have from most outlets. Despite the internet boom, most reporters were new to tech coverage, and almost none had expertise in security. *Wired News* was different. Its reporters knew about tech, security, and the Cult of the Dead Cow. So when mainstream reporters read *Wired News'* coverage of the Hong Kong Blondes, they assumed the publication had vetted its sources and knew what it was talking about.

cDc members had mixed first reactions to Laird's text file, but most were impressed. They believed the story because the details in his writing matched what he had told them before. But Kevin and cDc text-files editor Misha, whose reputations were more on the line, smelled something funny. Looking over a prepublication draft, Kevin wrote to the list: "The Blondie Wong interview is great. How much of that is real?" Laird replied: "It's three-quarters real and the rest is a buncha yang." Misha, charged with editing it, was more blunt in writing directly to Laird. "For the most part, he [Blondie] doesn't get some of your humorous turns of phrase and he himself speaks in a very formal, careful language, and then out of the blue in parts, he says stuff like, 'The guy's an idiot. I mean, if I want advice from the president about getting a blow job from a young girl, I'm all ears.' I gotta ask: Is this interview for real? Or did you write both parts?"

For all his doubts, Misha was backed into a corner. Laird had already given the file to *Wired News*, which had used it to publish its own story. "As leader of the Hong Kong Blondes hacking group, Wong has the credentials to back up his threats," Hesseldahl wrote. "The Hong Kong Blondes claim to have found significant security holes within Chinese government computer networks, particularly systems related to satellite communications." It would look awfully strange if cDc did not print its own scoop. Besides, Misha thought the piece might raise awareness, and he had been solidly behind media pranks in the past. He smoothed out Laird's interview and published it on the cDc site.

After the *Wired News* piece, Naomi Klein got in touch. The rising Canadian journalist saw the Toronto angle and was especially interested in China. Clinton had been working to normalize relations and de-emphasize human rights, and he had just conducted the first presidential visit to the nation since the Tiananmen massacre. "She thinks we're this righteous politicized hacking machine out for world peace or somethin'. . . . Anyway, we're gonna get a lot of miles outa this baby," Laird wrote to the group. He was right. Klein's wide-eyed write-up in the *Toronto Star* reported that "the Blondes are the hacker wing of China's pro-democracy movement, scattered around the world and forced underground after Tiananmen. On July 7, days after Bill Clinton returned from his trip to China, Blondie Wong, the pseudonymous director of the Blondes, met with Ruffin and the two went public with a new level in political hacking." Many other outlets picked up Klein's account. She would go on to write books including *No Logo: Taking Aim at the Brand Bullies*, which also quoted Blondie Wong.

The story seemed to have come from the future. It flew from site to site on the still-young web, a fantastic tale of accomplished, mysterious hackers aiding heroic human rights activists inside a totalitarian world power. Though no one besides Laird even claimed to have been in contact with Blondie—described variously as an astrophysicist and a currency trader—or to have any other corroboration, more stories followed, bolstered by the innocence of the reporters and the fact that cDc previously had established itself in the national media as an elite club of hacking gurus.

cDc was now ecstatic, and it used the Blondes to stake out moral high ground. As Kevin paced the Def Con stage to launch Back Orifice and, a year later, BO2k, he cited the Blondes as the prime example of what the group was fighting for. When Microsoft switched from dismissing Back Orifice as a toy to calling it dangerous and attacking cDc for releasing it, Luke sent out a press release tying the company to China and suggesting that "hacktivists" use Back Orifice to attack businesses in bed with the regime.

Was releasing Back Orifice to the public immoral? Microsoft would love for their customers to believe that we're the bad guys and that they—as vendors of a digital sieve—bear no responsibility whatever. But questions of morality are more often relative than absolute. So to make things easier, we'll frame our culture and actions against theirs, and let the public determine which one of us looks better in black. We'd like to ask Microsoft, or more to the point, we'd like to ask Bill Gates why he stood shoulder to shoulder in 1996 with China's president and head of the Communist Party to denounce any discussion of China's human rights record at the annual meeting of the United Nations Commission on Human Rights in Geneva? Was the decision to cozy up to the world's largest totalitarian state based on some superior moral position, or was it just more convenient to trample human decency underfoot and go for even more money? . . .

Now let's return to Back Orifice. Would it be immoral to use this tool for untoward purposes on Windows networks? Would it be immoral for Back Orifice to find its way to China and cause a lot of dry heaving in Microsoft's largest target market? Should hacktivists use Back Orifice as a form of protest against multinationals who share Microsoft's position of dollars before dignity? It's a short life and we're all going to be judged by our actions. So, whether or not we've done the right thing is a matter for history and human conscience to decide. But if the gods want to curse us for bringing fire down from the mountain, we'll take a seat with Prometheus and deal with the heat. At the end of the day, the CULT OF THE DEAD COW doesn't think that the world was meant to be a dark place.

The group worked carefully on media strategy, matching reporters with stories and members who were the most quotable. Laird urged them all to use the word *hacktivism*, no matter what the questions were about. "If ten different journalists all file stories with the same word in it, it [hacktivism] shoots into the common lexicon very quickly," he wrote to the mail list in July 1998. "Only get to their question after you've said what you want to say, then touch on theirs if it's worth answering, or just ignore it completely. . . . After this, you'll all be able to run for public office." The strategy worked beyond anything he could have imagined. In January 1999, a respected China-based writer for the *Los Angeles Times* wrote a front-page feature story about various ways the net was being used by democracy advocates in China. She cited cDc and the Blondes as hacktivist groups fighting the Great Firewall. Then she quoted both Oxblood and Blondie Wong's purported statements to him.

[x x]

Not all of the press fell for the Hong Kong Blondes saga. Back Orifice was a clear story, with lots of experts involved, public demonstrations, and major companies issuing warnings. The Hong Kong Blondes had nothing verifiable behind it. Without anything to go on other than Laird's word, most responsible publications didn't print anything. As more asked for access to Blondie Wong in order to go ahead with their own stories, Laird said he had disappeared. In December 1998, just before the *LA Times* story appeared, he wrote another account of their relationship. In cDc text file #361, he said he had met Wong at a party by accident three years earlier, that they had jointly come up with the structure of the Blondes within hours, and that Wong had recently moved to India, mainly to work with South Asian programmers.

Then Laird changed the subject, citing irrefutable evidence of rights abuses in China and touting open-source software as having better potential for improving lives there than Western governments or companies. He also named genuine Chinese activists, and he said that hacktivists could assist them in multiple ways. They could get Back Orifice into China for use against corrupt party officials, and they could help fight for attention. In a prescient forecast, he said that hacktivism was powerful and that it was largely about spreading knowledge in a new kind of conflict, "the information war where memes compete for mindshare and ratings replace body count."

Hackers and activists took notice of the Blondes story, and some defaced Chinese government websites. One group of US hackers, Legions of the Underground, in December 1999 called on allies to destroy networks in China and Iraq. Within days, cDc issued a joint statement with the L0pht, *Phrack*, and the long-political Chaos Computer Club of Germany in response. "Though we may agree with LoU that the atrocities in China and Iraq have got to stop, we do not agree with the methods they are advocating," the statement read. "One cannot legitimately hope to improve a nation's free access to information by working to disable its data networks. . . . If hackers are establishing themselves as a weapon, hacking in general will be seen as an act of war." The LoU, which had been split internally over the matter, took the warning seriously and decided to call off the attack.

cDc tried to redirect the energy in a more defensible way. Laird worked with some of the other politically inclined members to create a cDc offshoot dubbed Hacktivismo. It nursed several projects for evading censorship and communicating securely, though none seemed to reach critical mass. In the meantime, the leading technical talents of cDc were focusing more on their day jobs. Mudge and Christien Rioux, in particular, took the L0pht in a shocking new direction. They and the rest of the group arranged for it to be bought by a for-profit company and took in venture money to go fully professional. cDc software and news releases slowed, and the group presentations at the summer 2000 hacking conventions included spectacle and dry updates but little in the way of fresh tools, news, or inspiration.

Some hackers grumbled that Laird had ruined cDc by making it political, and a few raised serious questions about the Blondes. More than a decade later, Laird walked the tale halfway back in a post on Medium, saying he had never met anyone besides Blondie and had made up parts of the story to protect him. Laird continues to insist that at least Blondie was real. But journalist Hesseldahl gradually realized he had been hustled. Twenty years after the fact, he said this: "The stories led to some interesting and constructive conversations in academic and intellectual circles around how hackers and activists might help each other. If those conversations led somehow to some positive change in the world, that's great. But it doesn't excuse me for either story."

[x x]

In the spring of 2001, a British transplant to India stumbled onto Laird's story about Blondie Wong. Greg Walton had been a bit of a hippie when he left the UK treadmill for northern India, where he found dancing and distraction. As the web flourished over the next few years, Walton thought he could help as a human rights consultant and by developing websites for the Tibetans in Dharamsala, the de facto capital of the Tibetans in exile from China. He was taking a break in the office of a cultural institution, reading text files, when he found Laird's three-year-old story about Blondie Wong. Walton couldn't contain himself. Here was a Canadian technology activist conspiring with persecuted Chinese dissidents. The Tibetans needed help like that. Every day, barrages of trick emails and all manner of electronic subterfuge came at them from China, which was bent on discrediting the Dalai Lama and stopping him from leading ethnic Tibetans still inside its borders. Walton fired off an email to Laird, thanking him for his work and asking if he knew anyone who could help with the Tibetans' woeful security.

Laird asked what Walton had been doing, then promised to think about aiding the cause. First, though, he invited Walton to come to Las Vegas that summer and join a panel at Def Con about the need for more security in the service of human rights. cDc had decided to make a stronger run at the idea of hacktivism.

Walton made the trip to Vegas and spent hours hanging out at the Hard Rock Hotel, across the street from the conference, with Laird and a parade of other hackers. cDc had control of the panel and had made an inspired pick for the main talk: Patrick Ball, deputy director of a rights project at the American Association for the Advancement of Science. Ball was an ace programmer and one of the first to do exactly what Laird and the other cDc members had been talking about. After he had dropped out of graduate school in 1991 and traveled to El Salvador, Ball had moved to one troubled country after another, methodically drafting programs, installing protective cryptography, and compiling databases of some of the worst human rights horrors in the world. He presented at the Computers, Freedom, and Privacy conference in 1998 in Austin and debated crypto policy with an official at the Department of Justice who wanted back doors and weak encryption. It was there he met Cypherpunks mailing list sponsor and EFF cofounder John Gilmore and Phil Zimmerman, the inventor of Pretty Good Privacy (PGP) email encryption, who had also battled the federal government. "I ended up being friends with these people forever," Ball said.

In Las Vegas, Ball and the others were speaking in a hot tent pitched on a hotel roof, because no ballroom was big enough. It was the largest crowd Ball had ever spoken to, perhaps seven hundred people. In El Salvador alone, Ball told the enthralled Def Con audience, his team had recorded nine thousand witness accounts describing torture, kidnappings, and extrajudicial killings. He compiled one database with reported crimes against seventeen thousand victims and another with the careers of thousands of people in the military, then merged them to discover which officers kept appearing in the worst abuse cases. He found one hundred who stood out, each with more than one hundred apparent crimes at his hands or under his watch, and got them fired from their positions. Ball's technologically advanced crusade had since taken him to Haiti, Guatemala, South Africa, and Serbia, where he found incontrovertible proofs of genocide, drove out some of the worst offenders, and changed the history books.

Now Ball appealed to those in the crowd to help as well—by writing a letter a month for Amnesty International on behalf of political prisoners, by joining US interest groups fighting restrictions on internet and security research, or by donating programming time to efforts like cDc's Peekabooty, a privacy-protecting browser. "Hacktivism is finding ways to speak truth to power," Ball said. Laird and Walton then joined Ball in discussing Hacktivismo, the cDc offshoot with a new mission statement about human rights. Mimicking the form of a United Nations resolution, Hacktivismo's initial July 4, 2001, "Hacktivismo Declaration" cited Article 19 of the Universal Declaration of Human Rights, which held that everyone has a right to freedom of opinion and expression, including the right to receive information. Hacktivismo's opening volley declared in all capital letters: "STATE-SPONSORED CENSORSHIP OF THE INTERNET IS A SERIOUS FORM OF ORGANIZED AND SYSTEMATIC VIOLENCE AGAINST CITIZENS, IS INTENDED TO GENERATE CONFUSION AND XENOPHOBIA, AND IS A REPREHENSIBLE VIOLATION OF TRUST." Hacktivismo pledged to use tech to fight back. "WE WILL STUDY WAYS AND MEANS OF CIRCUMVENTING STATE SPONSORED CENSORSHIP OF THE INTERNET AND WILL IMPLEMENT TECHNOLOGIES TO CHALLENGE INFORMATION RIGHTS VIOLATIONS."

Laird had written the first draft in Kevin's Harlem apartment, calling it the "Harlem Declaration" while he thought up a permanent title. He crafted it carefully, with input from Luke, Misha, and others. Laird also consulted lawyers, including cDc member Glenn Kurtzrock, who was working as a prosecutor on Long Island, and Electronic Frontier Foundation contract lawyer Cindy Cohn, who would later lead that nonprofit. Cohn helped Laird to combine moral authority with UN legitimacy to reach as big an audience as possible without inviting condemnation by governments. A key idea was citing not just the Universal Declaration of Human Rights, which was nonbinding, but also the International Covenant on Civil and Political Rights, which was less well-known but had the force of a treaty. "I didn't write the 'Harlem Declaration' to preach to the converted," Laird said in explaining some of his choices to the others. "If that were the case I would have written something along the lines of 'Li Peng is a cocksucker who's out to destroy the internet.'"

In a public FAQ post, Laird and the other core Hacktivismo members wrote of the declaration: "The main purpose was to cite some internationally recognized documents that equate access to information with human and political rights; to state unequivocally that reasonable access to lawfully published material on the Internet is a basic human right; that we're disgusted with the political hypocrisy and corporate avarice that has created this situation; and, that we're stepping up to the plate and doing something about this."

The 2001 Def Con panel was a defining moment for those onstage and many in the audience. "It really spoke to me," Walton said. If Laird had before been an occasional agitator, a dilettante provocateur, Ball was the real deal, and he struck a nerve with the Def Con hackers. The following year, he testified at the war crimes trial of Serbian dictator Slobodan Milošević at The Hague. Milošević, acting as his own attorney, sought to discredit Ball and asked him on cross-examination: "Mr. Ball, did you in the—the advisory board of the hacktivism group of international computer hackers, are you in the management board of that group which is known as the 'Dead Cow Cult'?" Ball said he merely advised cDc in its "efforts to try to help young computer programmers move away from illegal activities and direct their activities toward productive and legal activities," including human rights work.

Walton kept in touch with Laird and helped him get a job organizing a conference in Dharamsala on wireless technologies that could expand the reach of the Tibetan monks while keeping them secure. Laird brought over technology luminaries who drew more people to the conference and inspired others to come help the Tibetan community. Walton, meanwhile, became far more sophisticated about the nature of the threats to the Tibetans, and he deepened his commitment to protecting those he was serving. Western intelligence, meanwhile, kept a close watch on the Tibetans because better defenses in Tibet meant that the Chinese would try more sophisticated attacks. Once the game got fancier, they knew that what the Chinese threw at the Tibetans today would be coming at big US defense contractors like Lockheed Martin tomorrow. Walton himself kept agents up-to-date on his efforts. "If you have the same malware samples reporting back to a command-and-control server that is attacking defense contractors and nunneries in Dharamsala, that's a pretty good indication" that China is driving it, Walton said. "So I shared with a couple national intelligence agencies. Laird was certainly well connected to all kinds of people too."

Actually, Laird was more connected to the intelligence establishment than even most of cDc realized. Before he picked Ball to speak at the Def Con panel that gave the greatest prominence to the idea of hacktivism, he had floated the idea of having James Mulvenon on the panel instead. Mulvenon, an intelligence contractor at RAND, was a China specialist focused on helping US intelligence, not the oppressed citizens of China. The two had connected after the Hong Kong Blondes articles, when Mulvenon was looking for any evidence of Chinese dissidents harnessing the net. But Chinese internet accounts were still rare in those days. Mulvenon had traveled to the mainland and looked deeply into hacking groups. The only ones he found with any political awareness at all, like the one called the Honker Union, were patriotic, often acting at the direction of the government.

Mulvenon had found nothing like what Laird wrote about. But his idea was so appealing that people within US intelligence promoted it as a goal to strive toward. "The Hong Kong Blondes were part of a story line that inspired people to believe it was a moral mission," said a US intelligence official who came to know Laird. Under President Barack Obama, the American government would take up the cause on its own, distributing tools for uncensored internet connections through Hillary Clinton's State Department. The program, informally known as "internet in a box," was championed by Clinton's innovation advisor, Alec Ross, who declined to say which countries it reached.

In the meantime, Laird spoke often to Mulvenon, briefing him in person on at least one cDc project for circumventing censors. That was the Peekabooty plan to route web page requests through strangers' browsers to obfuscate who was viewing what. Laird and Mudge were not the only ones in cDc close to Western intelligence. Member Adam O'Donnell, known as Javaman, also worked on a CIA project to reverse engineer the Great Firewall of China, the nickname for the system China uses to control internet traffic into and out of the country, and to figure out how to get content inside. Brought into cDc in 2004, Adam was the son of two supermarket employees, one of whom later earned a doctorate in education. After attending Philadelphia's magnet Central High School and heading to college, he interned at Lucent Technologies, the old Bell Labs. He also went to *2600* meetings in New York. People there took him to Boston to meet the L0pht guys. Later, in California, he worked for anti-spam company Cloudmark, begun by fellow w00w00 member Jordan Ritter, before cofounding security firm Immunet in 2009.

The reverse proxies Adam built allowed people in the US to make it appear that they were in China, mainly so they could see what the Firewall was blocking. It also made it easier for those Americans to share information with people who really were in China. Finally, it put the CIA in a better position to monitor dissident traffic or hack Chinese targets without raising alarms about foreign intrusion.

It seemed like a decent way to help people in China and make a bit of money off the government, but it occasionally gave Adam an unpleasant feeling. Once he had to pick up a payment at a dive bar in Washington's Dupont Circle. The men he met didn't say much, but they handed over a Chipotle bag with $20,000 in cash inside. When Adam asked if they needed a receipt, they just laughed. "Not for anything less than a hundred K," one said.

While some in cDc tried using the government to advance the cause of internet freedom and security, others were hoping to do good by riding the internet boom.

CHAPTER 8

MUCH @STAKE

THE BEGINNING OF the year 2000 brought the absolute peak of the
dot-com bubble. Though they were by no means a driver of the
stock-market insanity or the venture capital greed behind it, digital security companies benefited from being hired by the likes of AOL and
Yahoo, computing and e-commerce firms, and others. Some of the companies employed a handful of talented penetration testers, who would
break into clients with permission and then advise them on how to fix
the holes they came through. Giant consulting firms with employees of
mixed abilities had a bigger presence. Then there were the major antivirus companies like Symantec and McAfee. Their products were better
than nothing, and their business models raked in cash. The companies
charged annual fees to consumers and businesses and blocked what
viruses they could. When a client got infected anyway, the companies
added the new signature to the detection database so the same virus
wouldn't hit the next guy—unless the virus changed slightly between
infections, that is. Unfortunately, making minor changes to a virus was
trivial for hackers targeting a specific victim, and soon enough such
changes became an automated part of broader attacks.

Companies earned good money with defense, but overall they
failed to make customers safe. On the contrary, as businesses bolted
together different software, hardware, and networks, security actually
got worse. Because every program of any size has critical flaws that

can be leveraged by an attacker, greater complexity aided hackers and handicapped defenders. But the software suppliers lacked the incentives that drove ordinary manufacturers to make safer products. The software companies had convinced the courts that product-liability laws did not apply to them. Technically, they licensed their products instead of selling them, and they forced users to waive the right to sue at the moment of installation. The biggest customers could try demanding assistance via service agreements or code audits. But even if customers won the right to examine the code for flaws, they had no right to warn other customers about what they found. More fundamentally, most major products had few good alternatives, and they all had flaws. Not even the biggest companies shopped for software mainly on the basis of security. At best, they encouraged employees to use and contribute to open-source projects like Linux. That helped in the server market but posed little threat to desktop operating systems, let alone the applications that ran on them.

Barring some executive-branch, legislative, or courtroom surprise, the L0pht crew figured the next best way to improve the world's security would be convincing the biggest software makers to do the right thing, even if they didn't have to. Public embarrassment, led by the Cult of the Dead Cow, had done more than anything else to persuade Microsoft to take security more seriously. But Microsoft was just one company, and shaming businesses brought the L0pht no cash. With just a little income from selling tools like password crackers, the L0pht couldn't scale. So Mudge and some of the others wondered if they could somehow get invited inside more software companies to at least make the bad stuff better. They could also consult with big banks and other customers, giving them ammunition to demand better software from suppliers. Enough new business and they could hire more hackers. If the L0pht did it right, they could work with both buyers and sellers and protect hundreds of millions of people.

Mudge wasn't sure the rest of the crew would see things the same way, but he didn't want to keep going the way things were. He, Christien Rioux, and Chris Wysopal were writing most of the L0pht programs that earned money—tools to scan networks, crack passwords, and so on. But that meant those three had to keep laboring to improve those programs even if they wanted to research something new. Tired of the burden, Mudge suggested getting outside investment from a venture capital firm, so they could all do what they enjoyed. Though he knew it would offend some purists who hacked for hacking's sake, not for money, Wysopal reasoned that hackerdom would be much better off with them getting paid to tinker. "Maybe it was something that was impossible to do, and we had wishful thinking that we could figure it out," he said. To Mudge, it was about distribution in the Kevin Wheeler sense, getting the word out about how unsafe things were and how to improve them. "We were the best garage band in the world at that time. And the only people who know you are the people on your block and maybe their friends," Mudge said. "So you take money from a record label. It comes with baggage, but the message gets out further."

The L0pht had no shortage of outside interest. A logical early contender was Cambridge Technology Partners. It was a security consulting group with some credibility that had just been featured on *20/20*, in a segment where Cambridge hacker Yobie Benjamin and others broke into a major unnamed bank on camera in a penetration test. When the L0pht met with Cambridge Technology, the members suggested that the company hire them for a penetration test against it. That way, Mudge said, the executives would know the L0pht's capabilities. In agreeing, Cambridge made a fatal mistake. After the last of the legal authorizations was signed, Joe Grand went straight for the executives' voice mail and tried the most obvious four-digit codes to listen to the messages: 1234, 1111, 4321. In short order, they knew what Cambridge was going to offer to buy the L0pht, what its best offer would be if the first offer got rejected, and, most awkwardly, what the executives thought of the L0pht's members. They really only wanted Mudge, Christien, and Wysopal. That was infuriating, but the discovery also gave them a license to have fun. The L0pht went back into negotiations with unusual demands, asking for a Winnebago like the guys in the hacking movie *Sneakers*. Then they turned over their report on the pen test. They weren't mean enough to include quotes from the voice mails, but it was obvious what had happened. They never heard from Cambridge again.

It got better with an approach from Battery Ventures, an established venture capital firm. Battery had just backed a fledgling start-up called @stake. @stake had hired Luke Benfey's old housemate Dave Goldsmith from Cambridge Technology and Window Snyder as well. They agreed to a $10 million deal that folded the L0pht into @stake when it closed in January 2000. Around then, overexcited public relations people told media the real names of Mudge, Christien, and Wysopal. They tried, too late, to claw the information back. And yet, the world didn't end. Professionals were brought in as the top executives, leaving the old L0pht crew free to continue doing their research. Hackers had such admiration for the L0pht that @stake pulled some of NSA's best to the private sector, and the new company became an odd marriage of security brains and money.

But the culture of the unkempt rebels in the rank and file clashed with that of the suits making sales pitches and controlling the budget. Sketchy pasts and big personalities abounded. Some employees missed a major customer meeting because they had been up all night doing drugs. Other meetings should have been missed but weren't: one L0pht veteran was having sex with a prostitute in the office when her rear end knocked into a phone and joined them to a conference call with a customer's CEO. And later, a former employee was jailed for playing a role in one of the largest thefts of credit card numbers ever detected.

More subtle issues also surfaced. Would @stake continue the L0pht's practice of issuing advisories about dangerous bugs? Or would it only do that about companies it did not work for as a consultant? If it wouldn't embarrass a company that was paying it, that could get dangerously close to extortion: "Hire us and we'll shut up about your product." Though @stake continued the tradition of coordinated disclosure that the L0pht had pioneered, its policies were impure. A bug found in a noncustomer's software—or found off the job in a customer's wares—could be disclosed, but it could also be used for business development. Customer bugs found during an engagement were kept quiet.

@stake needed to sort out its disclosure policies quickly, because none other than Microsoft hired it for major work at the company. Despite the past antagonism, the @stake crew made a huge positive impression at Microsoft. Like a task force of star detectives, they possessed a sixth sense about where problems hid in the code. They followed connections from one product to another, and they looked at work patterns as well. Several versions of Windows had substantially better security because of @stake, and in 2002 Bill Gates released a memo declaring that security was now the company's top priority.

Microsoft soon hired Snyder and other @stake veterans in-house. Snyder would stay three years. In the beginning, the company had no single person responsible for security issues in upcoming versions of the operating system. Snyder raised her hand. She still had to fight for things that cost money, like delaying a release to fix bugs. Arguing with the managers of a version about to go "gold" for general release, she said Microsoft should first plug two medium-level vulnerabilities, because someone outside would find them and build on those flaws to make something more dangerous. She lost the vote and a few days later was proven right. After that, the other managers stopped arguing with her. Snyder brought in many of the best outside security consultants, and she was responsible for Windows XP Service Pack 2, which dramatically improved the company's posture. Snyder also helped expose isolated executives to outside researchers by creating the BlueHat security conferences, at which hackers spoke for an audience of Microsoft employees.

@stake staff and veterans entered new territory in other ways as well, including by publishing research that brought unintended consequences. David Litchfield, a Scot on his way to becoming the world's best-known database security expert, was gone from @stake and testing the security of an SQL database for a German bank when he had a harder time than usual breaking in. Litchfield tried sending various single bytes and found one that crashed the system. That led to more experimenting and then a short program that might be able to take control of the database. More digging found a surefire way to exploit a similar flaw. Litchfield warned Microsoft and asked if he could present a talk on the matter at Black Hat, the more professional version of Def Con that now ran just before it on the calendar. Microsoft had no problem with that; it would have a patch ready by then. Litchfield's talk included sample code, and he warned everyone to install the patch. Six months later, an unknown coder released SQL Slammer, a self-replicating worm that shut down large parts of the internet in 2003. Only about 10 percent of machines had been patched, Litchfield guessed. Certainly many of the companies would not have been hurt if he had not published actual code. So Litchfield resolved only to describe such dangerous flaws in the future, not release proof-of-concept code, unless he could be sure nearly everyone had patched.

@stake Chief Technology Officer Dan Geer further tested the company's willingness to speak the truth by cowriting a 2003 paper arguing that Microsoft's monopoly was bad for security. Geer's team said that Microsoft's dominance made it worthwhile for hackers to focus on finding its weaknesses, because they would provide a golden key that would get them in almost everywhere. It was true, but it was also a provocation, and it came just as Microsoft's court-certified monopoly was finally waning under pressure from a rejuvenated Apple. @stake unceremoniously fired Geer by press release.

The one truly insurmountable problem for @stake was venture capital math. Battery Ventures knew that most of the companies it invested in would fail, so it concentrated on the ones it thought could potentially deliver "100x" returns, the home runs. But the money coming in to @stake was in consulting, and the company could never have produced those kinds of returns. To satisfy its investors, @stake would have had to grow as big as one of the largest management-consulting firms. @stake limped on through its 2004 sale to Symantec, which gradually absorbed it.

[x x]

The @stake story was a strange shotgun union of two powerful and growing forces: venture capital and hacking. In its short arc, @stake

established an enormously important precedent for security: that outsiders could go into big companies and make the systems and products
there safer. Perhaps more importantly, @stake hackers dispersed and
founded many more companies in the next few years, and they became
security executives at Microsoft, Apple, Google, and Facebook.

But those same years revealed psychological fragmentation in
the movement along with the physical diaspora. The cDc of Def Cons
1998 through 2001 had ridden the crest of a wave of hacker sensibil-
ity. Each year the crowds grew in number, young, irreverent, and on
the cusp of mass recognition, if not big money. That short period was
as important for technology culture as the Summer of Love, in 1967
San Francisco, was for the hippies. Laird Brown's hacktivism panel in
the summer of 2001 set a high-water mark for that kind of enthusiasm,
for open-source, idealistic efforts to protect people even from their own
government.

But any youthful protest ethic faces a challenge when its adherents
need to find jobs and pay their bills. That concern increased in 2001,
one year into the great bust that followed the dot-com boom. Not everyone could get a job with @stake or other boutiques. But it was a second,
more direct blow that scattered young hackers in different directions
for many years: the terrorist attacks on the World Trade Center and the
Pentagon.

Those driven primarily by money were already paying less attention
to ethical quests, such as the fun and games in keeping Microsoft hon-
est. Now, in the months after the 9/11 attacks, those driven largely by
causes also had a strong contender for their attention: rallying against
the worst attack on American soil since Pearl Harbor. This was true for
rank-and-file hackers, who took assignments from the military or intelligence agencies, and even cDc's top minds, including Mudge.

Mudge had instant credibility, since he had taught government
agents and they used his tools. Government red team penetration-test
leader Matt Devost, who had covered cDc in a report given to a presidential commission on infrastructure protection, used L0pht tools to
break into government networks. Spies loved Back Orifice and BO2k

because if they left traces behind, nothing would prove US government responsibility.

Two years before 9/11, an intelligence contractor I will call Rodriguez was in Beijing when NATO forces in the disintegrating state of Yugoslavia dropped five US bombs on the Chinese embassy in Belgrade, killing three. Washington rapidly apologized for what it said had been a mistake in targeting, but the Chinese were furious. In a nationally televised address, then Chinese vice president Hu Jintao condemned the bombing as "barbaric" and criminal. Tens of thousands of protestors flowed into the streets, throwing rocks and pressing up against the gates of the American embassy in Beijing and consulates in other cities.

The US needed to know what the angry crowds would do next, but the embassy staffers were trapped inside their buildings. Rodriguez, working in China as a private citizen, could still move around. He checked with a friend on the China desk of the CIA and asked how he could help. The analyst told Rodriguez to go find out what was happening and then get to an internet café to see if he could file a report from there. Once inside an internet café, Rodriguez called again for advice on transmitting something without it getting caught in China's dragnet on international communications. The analyst asked for the street address of the café. When Rodriguez told him exactly where he was, the analyst laughed. "No problem, you don't have to send anything," he explained. "Back Orifice is on all of those machines." To signal where he wanted Rodriguez to sit, he remotely ejected the CD tray from one machine. Then he read everything Rodriguez wrote as he typed out the best on-the-ground reporting from Beijing. Rodriguez erased what he had typed and walked out, leaving no record of the writing.

Even before 9/11, Mudge had been talking to Richard Clarke and others at the National Security Council. Often, Mudge argued for privacy. The government had wanted to put location tracking in every cell phone as part of Enhanced 911 services, for example. Mudge told the NSC that the privacy invasion was unnecessary, that information from cell phone towers would be good enough for any serious official need.

One day in February 2000, after a rash of denial-of-service attacks that bombarded big websites with garbage traffic so that regular users couldn't connect, Richard Clarke brought Mudge into a White House meeting with President Bill Clinton and a bunch of CEOs. "It was, I think, the first meeting in history of a president meeting people over a cyber incident," said Clarke, who had organized it to show White House responsibility on the issue and build the case internally for more government oversight. After answering Clinton's questions on what was fixable and what wasn't, the guests walked out of the office. The CEOs saw the reporters waiting and prepared their most quotable platitudes. Instead, the press swarmed Mudge, as even those who didn't know him assumed that the guy who resembled a Megadeth guitarist was a hacker meeting with the president for good reason. "Of course Mudge stole the show," Clarke said.

But in order to be taken seriously, Mudge had to tell the truth. Once, an NSC staffer brought him in and asked what he knew about a long list of terrorists and other threats. What did he know about Osama bin Laden? About the group behind the sarin attack in the Japanese subway? About the Hong Kong Blondes?

At that one, the blood drained from Mudge's face. "What do you mean?" he asked.

"We've been informed it's a small, subversive group inside China that's helping dissidents with encrypted communications," the staffer replied.

"I've heard of them," Mudge offered.

"What can you tell us?" the staffer persisted.

Mudge figured the government hadn't put a lot of resources into the goose chase because signals intelligence and other sources would have turned up nothing and convinced seasoned professionals that it was a red herring. But he didn't want the country to waste any energy that could go toward supporting real people in need.

He shrugged and looked straight at the staffer. "We made them up," Mudge admitted.

After 9/11, Mudge went into overdrive. President Bush was warned that a cyberattack would have been worse than the planes, and he listened. Mudge then started exploring what a "lone wolf" terrorist hacker could do. "I'm finding ways to take down large swaths of critical infrastructure. The foundation was all sand. That rattled me," Mudge said. Looking into the abyss exacerbated Mudge's severe anxiety, his tendencies toward escapist excess, and his post-traumatic stress disorder, which had its roots in a violent pre-L0pht mugging that had injured his brain. He went into a spiral and eventually broke down. "Ultimately, I just cracked a bit," Mudge said. He spent days in a psychiatric ward. (Anxiety and burnout in the face of the near-impossible, high-stakes task of defending networks was not yet recognized as a major industry problem, as it would be a decade later.) Unfortunately, some of Mudge's treatment compounded the situation. As is the case with a minority of patients, his antianxiety medications had the opposite of the intended effect. Eventually, Mudge fired his doctors, experimented with different medications and therapy, and worked his way back to strong functionality. But when he returned to @stake after many months, it was too fractious and uninspiring for him to be enthusiastic about reclaiming his post. The dot-com bust had forced layoffs of L0pht originals while managers were drawing huge salaries. The emphasis was on the wrong things.

Outside of @stake, hackers began disappearing from the scene for six months or more. When they came back, they said they couldn't talk about what they had been doing. Those who went to work for the intelligence agencies or the Pentagon, temporarily or permanently, included many of the very best hackers around, including a few present or former cDc members and many of their friends in the Ninja Strike Force. They wanted to protect their country or to punish Al-Qaeda, and in many cases they got to work on interesting projects. But many of them would not have passed the background investigations required for top secret clearances. To get around that problem, a large number worked for contractors or subcontractors. One way or another, a lot of their work went into play in Afghanistan and Iraq.

Cult of the Dead Cow distributed text files to as many bulletin boards as wanted them, and it promoted the home boards of members, including Jesse Dryden's K0de Abode.

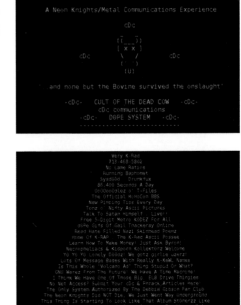

In 1990, Texas hacker Jesse Dryden created HoHoCon, the first hacking conference to invite law enforcement and the media. Courtesy Rodney Palmer

At SummerCon in 1992, the Cult of Dead Cow's Misha Kubecka and Dan MacMillan traded stories with *Phrack* editor Craig Neidorf, whose misguided hacking prosecution inspired the founding of the Electronic Frontier Foundation. Courtesy Colin Campbell

The Cult of the Dead Cow panel at New York's Hackers on Planet Earth conference in 1997 included Carrie Campbell, Laird Brown, Luke Benfey, Sam Anthony, and Peiter "Mudge" Zatko. Speaking as Oxblood Ruffin, Laird offered the first public claims about the Hong Kong Blondes. Courtesy Abby Fichtner

At a party one evening during HOPE 1997, Carrie Campbell introduced Psychedelic Warlord to newer members of the cDc. Courtesy Danny Dulai

Seven members of the L0pht testified before Congress in May 1998 under their handles, including in the middle seats "Kingpin" Joe Grand, cDc "hacker laureate" Mudge Zatko, and Ninja Strike Force member Chris Wysopal, known as "Weld Pond." Courtesy Douglas Graham/*Congressional Quarterly* via Getty Images

As Grandmaster Ratte, Kevin Wheeler commanded the Def Con stage for the launches of Windows trojans Back Orifice in 1998 and BO2k in 1999. Courtesy Abby Fichtner

---

FEDERAL BUREAU OF INVESTIGATION

Precedence: ROUTINE   **Date:** 07/22/1999

To: Atlanta

From: Atlanta
       Contact: X6185   b7E

Approved By: b6
Drafted By: b7C

Case ID #: 288A-AT-87389 (Closed)

Title: UNSUB (S);
     AKA: DETH VEGETABLE;
     NET NINJA;
     DBA: CULT OF THE DEAD COW;
     MINDSPRING ENTERPRISES - VICTIM;
     INTRUSION - INFO SYSTEMS.
     IDENTITY THEFT;
     CONSPIRACY

Synopsis: It is recommend that the above captioned case be closed.

Details: The above captioned group has been in existence since 1984. At lasts years DEFCON VI HACKER CONVENTION, the group released a product called "BACK ORIFICE". At this years convention which was held on July 10, 1999 in Las Vegas, Nevada, the group released a newer version called "BACK ORIFICE 2000". The product will be made available as a free download.

     BACK ORIFICE 2000 will allow an individual to gain access to a persons computer and retrieve system information including current user, cpu type, windows version, memory usage, mounted disks, drive information, screen saver password, and passwords cached by users. It will also allow the individual to have file system control, process control, registry control, network control, multimedia control, packet redirection, sniffing, application redirection and HTTP server.

     BACK ORIFICE 2000 will reportedly include several features not found in the original version, including windows NT compatibility.

     BACK ORIFICE 2000 will likely be used in a selective or targeted manner similar to previous network security exploits.

The FBI investigated both Back Orifice and BO2k but found no criminal activity by Luke "Deth Vegetable" Benfey or cDc. Courtesy of cDc

During the L0pht's 2000 farewell party in Boston, Christien Rioux caught up with Laird Brown under a banner filled with handles. Courtesy Abby Fichtner

In early 2000, President Bill Clinton held the first White House meeting on cybersecurity, speaking with Mudge Zatko and internet co-inventor Vint Cerf. Courtesy of the White House

Two early cDc Ninja Strike Force members, Limor Fried and Window Snyder. One of the initial leaders of the maker movement, Fried became the first female engineer on the cover of *Wired* magazine. Snyder played critical security roles at Microsoft and Apple, where she set the stage for the company to appeal a court order to break into a terrorist's iPhone. Courtesy Danny Dulai

German hacker Kemal Akman was a key addition to Laird Brown's Hacktivismo spinoff from cDc. But then Akman brought in the man who invented FinFisher spyware, still used by repressive regimes against dissidents. Courtesy Declan McCullagh

Jacob Appelbaum became one of the last close aides to WikiLeaks founder Julian Assange, with whom he spoke at a 2011 press conference touting the release of new files. Both would be accused of sexual misconduct. Still from "Wikileaks Press Conference," uploaded to YouTube December 1, 2011, by Rima Amin

Cult of the Dead Cow founders Bill Brown and Kevin Wheeler, known to the outside world as Franken Gibe and Grandmaster Ratte, met as middle-school students in Lubbock, Texas. Courtesy Kevin Wheeler

Security engineer and longtime Cult of the Dead Cow member Adam O'Donnell and Facebook security chief Alex Stamos hosted an early fundraiser for Senate candidate Beto O'Rourke of Texas during the fall of 2017. Courtesy Joseph Menn

Original cDc patch, from early hacking conferences. Courtesy Misha Kubecka

Some hackers felt great fulfillment in government service. Serving the government in the wake of the terror attacks gave them a chance to fit in when they hadn't before, united by a common cause. But for too many of this cohort, what started with moral clarity ended in the realization that morality can fall apart when governments battle governments. That was the case with a cDc Ninja Strike Force member I will call Stevens. As Al-Qaeda gained notoriety and recruits from the destruction, the US Joint Special Operations Command, or JSOC, stepped up the hiring of American hackers like Stevens. Some operatives installed keyloggers in internet cafés in Iraq, allowing supervisors to see when a target signed in to monitored email accounts. Then the squad would track the target physically as he left and kill him.

After 9/11, the military flew Stevens to another country and assigned him to do everything geek, from setting up servers to breaking into the phones of captured terrorism suspects. Though he was a tech specialist, the small teams were close, and members would substitute for each other when needed. Sometimes things went wrong, and decisions made on the ground called for him to do things he had not been trained in or prepared for mentally. "We did bad things to people," he said years later, still dealing with the trauma.

Others had similar experiences. A longtime presenter at hacking and intelligence community gatherings, former clergyman Richard Thieme, gave talks about the burdens of protecting secrets that should be known and about the guilt suffered by people made to carry out immoral orders. After he asked people to send in their stories, some listeners provided accounts like Stevens's. "It occurs to me how severely the trajectory of my own career has taken me from idealistic anarchist, to corporate stooge, to ambitious entrepreneur, to military/intelligence/defense/law enforcement adviser," wrote one. "Many cyber guys started out somewhere completely different and then somehow found themselves in the center of the military-industrial complex in ways they would never have been prepared for." Once there, the difficulty in keeping secrets is "potentially more extreme because the psychological

make-up and life-story of the cyber guy would not have prepared him
for it."

Wrote another:

> When one joins an intelligence service at the start of one's career,
> one is involved in low level, apprentice-like, tasks and assignments
> usually far removed from traumatic action or profound moral consid-
> erations, much less decisions. In the course of a career such actions/
> decisions slowly grow into being, almost imperceptibly for many peo-
> ple. One may suddenly "awake" to where one is and realize that he/
> she had not been prepared for this, and also realize that one is now
> deeply into the situation, perhaps well beyond a point that one would
> have stepped into if it had been presented from the start. If this is
> the case, it's too late to turn back.

When you are on the ground, Thieme said, "the rules people think
they live by are out the window." People who score too high on morals
tests are rejected by intelligence services, he said, because a conscien-
tious whistle-blower is even more dangerous than an enemy mole.

[x x]

Working for a contractor was just one way hackers with criminal histories and dicey connections could do business with the feds. Without
even going to that much effort, they could perform something close to
pure security research for cash. Penetrating many of the most valuable
and difficult intelligence targets required the government to have secret
knowledge of a software flaw. Those flaws had to be severe enough to
allow external hackers to gain control over a targeted machine. And
they also needed an exploit program that would take advantage of the
flaw and install software for spying. The National Security Agency, and
to a lesser extent other parts of the military and the CIA, had been
quietly developing storehouses of such flaws for years, along with the

exploits to take advantage of them. But both needed to be continually replenished. Once exploits were used, they could be discovered. Even if they weren't, it was dangerous to use the same technique elsewhere, because the target or a third country could realize the attacks were connected and draw conclusions about who was responsible.

As the American government ramped up its spying efforts after 9/11, it needed to discover new vulnerabilities that would enable digital break-ins. In the trade, these were often called "zero-days," because the software maker and its customers had zero days of warning that they needed to fix the flaw. A ten-day flaw is less dangerous because companies have more time to develop and distribute a patch, and customers are more likely to apply it. The increased demand for zero-days drove up prices.

After the dollars multiplied, hackers who had the strongest skills in finding bugs that others could not—on their own or with specialized tools—could now make a living doing nothing but this. And then they had to choose. They could sell directly to a government contractor and hope that the flaw would be used in pursuit of a target they personally disliked. They could sell to a contractor and decide not to care what it was used for. Or they could sell to a broker who would then control where it went. Some brokers claimed they sold only to Western governments. Sometimes that was true. Those who said nothing at all about their clients paid the most. For the first time, it was relatively straightforward for the absolute best hackers to pick an ethical stance and then charge accordingly.

It was in no one's interest to describe this market. The government's role was classified as secret. The contractors were likewise bound to secrecy. The brokers' clients did not want attention being paid to their supply chain. And the majority of hackers did not want to announce themselves as mercenaries or paint a target on themselves for other hackers or governments that might be interested in hacking them for an easy zero-day harvest. So the gray trade grew, driven by useful rumors at Def Con and elsewhere, and stayed out of public sight for a decade.

The first mainstream articles on the zero-day business appeared not long before Edward Snowden disclosed that it was a fundamental part of US government practice, in 2013.

As offensive capabilities boomed, defense floundered. Firms like @stake tried to protect the biggest companies and, more importantly, get the biggest software makers to improve their products. But just like the government, the criminal world had discovered hacking in a big way. Modest improvements in security blacklisted addresses that were sending the most spam. That prompted spammers to hire virus writers to capture thousands of clean computers that they could use to evade the spam blocks. And once they had those robot networks, known as "botnets," they decided to see what else they could do with them. From 2003 on, organized criminals, a preponderance of them in Russia and Ukraine, were responsible for most of the serious problems with computers in America. In an easy add-on to their business, the botnet operators used their networks' captive machines to launch denial-of-service attacks that rendered websites unreachable, demanding extortion payments via Western Union to stop. They also harvested online banking credentials from unsuspecting owners so they could drain their balances. And when they ran out of ideas, they rented out their botnets to strangers who could try other tricks. On top of all that, international espionage was kicking into higher gear, sometimes with allies in the criminal world aiding officials in their quests.

Out of @stake came fodder for both offense and defense. On offense, Mudge pulled out of his tailspin and worked at a small security company, then returned to BBN for six years as technical director for intelligence agency projects. His @stake colleague and NSA veteran Dave Aitel started Immunity Inc., selling offensive tool kits used by governments and corporations for testing, and for spying as well. He also sold zero-days and admitted it in the press, which was seldom done in those days due to ethical concerns and fear of follow-up questions about which customers were doing what with the information. Aitel argued that others would find the same vulnerabilities and that there was no reason to give his information to the vendors and let them take advantage of his work for free. From the defender's perspective, "once you accept that there are bugs you don't know about that other people do, it's not about when someone releases a vulnerability, it's about what secondary protections you have," Aitel said, recommending intrusion-detection tools, updated operating systems, and restrictive settings that prevent unneeded activity.

A London @stake alum moved in above a brothel in Thailand, assumed the handle the Grugq, and became the most famous broker of zero-days in the world. Rob Beck, who had done a stint with @stake between Microsoft jobs, moved to Phoenix and joined Ninja Strike Force luminary Val Smith at a boutique offensive shop that worked with both government agencies and companies. Careful thought went into what tasks they took on and for whom. "We were pirates, not mercenaries," Beck said. "Pirates have a code." They rejected illegal jobs and those that would have backfired on the customer. One of @stake's main grown-ups, CEO Chris Darby, in 2006 became CEO of In-Q-Tel, the CIA-backed venture capital firm in Silicon Valley, and Dan Geer joined as chief information security officer even without an agency clearance. Darby later chaired Endgame, a defense contractor that sold millions of dollars' worth of zero-days to the government before exiting the business after its exposure by hackers in 2011.

On defense, Christien Rioux and Wysopal started Veracode, which analyzed programs for flaws using an automated system dreamed up by Christien in order to make his regular work easier. After Microsoft, Window Snyder went to Apple. Apple's software had fewer holes than Microsoft's, but its customers were more valuable, since they tended to have more money. Snyder looked at the criminal ecosystem for choke-points where she could make fraud more difficult. One of her innovations was to require a developer certificate, which cost $100, to install anything on an iPhone. It wasn't a lot of money, but it was enough of a speed bump that it became economically unviable for criminals to ship malware in the same way.

Going deeper, Snyder argued that criminals would target Apple users less if the company held less data about them. But more data also made for a seamless user experience, a dominant theme at Apple, and executives kept pressing Snyder for evidence that consumers cared. "It was made easier when people started freaking out about Snowden," Snyder said. "When people really understand it, they care." In large part due to Snyder, Apple implemented new techniques that rendered iPhones impenetrable to police and to Apple itself, to the great frustration of the FBI. It was the first major technology company to declare that it had to consider itself a potential adversary to its customers, a real breakthrough in threat modeling. Still later, Snyder landed in a senior security job at top chipmaker Intel.

David Litchfield feuded publicly with Oracle over the database giant's inflated claims of security. He went on to increasingly senior security jobs at Google and Apple. @stake's Katie Moussouris, a friend to cDc, stayed on at new owner Symantec and then moved to Microsoft, where she got the company to join other software providers in paying bounties to hackers who found and responsibly reported significant flaws. Moussouris later struck out on her own and brought coordinated-disclosure programs to many other organizations, including the Department of Defense. She also worked tirelessly to stop penetration-testing tools from being subject to international arms-control agreements.

Private ethics debates turned heated and even escalated into intramural hacking. Some highly skilled hackers who found zero-days and kept them condemned the movement toward greater disclosure. Under the banner of Antisec, for "antisecurity," the most enthusiastic of this lot targeted companies, mailing lists, and individuals who released exploit code. In the beginning they argued that giving out exploits empowered no-talent script kiddies, like those who might have been responsible for SQL Slammer. But some of them simply didn't want extra competition. The mantle was taken up by hacker Stephen Watt and a group calling itself the Phrack High Council, which made the Antisec movement pro-criminal. Watt later did time for providing a sniffer, which recorded all data traversing a network, to Albert Gonzalez, one of the most notorious American criminal hackers. In a 2008 *Phrack* profile that used his handle only, Watt bragged about starting Project Mayhem, which included hacks against prominent white hats. "We all had a lot of fun," Watt said. Later on, the Antisec mission would be taken up by a new breed of hacktivists.

[x x]

Ted Julian, who had started as @stake marketing head before it merged with the L0pht, cofounded a company called Arbor Networks with University of Michigan open-source contributor and old-school w00w00 hacker Dug Song; their company became a major force in stopping denial-of-service attacks and heading off self-replicating worms for commercial and government clients. Song would later found Duo Security and spread vital two-factor authentication to giant firms like Google and to midsize companies as well.

Song got to know cDc files and then members online before being wowed in person by the Back Orifice release. In 1999, he put out dsniff, a tool for capturing passwords and other network traffic. While Arbor was mulling more work for the government, Song quietly developed a new sniffer that captured deeper data. He planned to show it off for Microsoft executives at Window Snyder's first BlueHat conference in 2004. Song went and talked about his improved sniffer, which analyzed instant-message contacts and documents and did full transcriptions of voice over IP calls, such as those on Skype. He produced a dossier on Microsoft employees as part of the demonstration. Then he decided the danger of such a surveillance tool outweighed the security benefit of catching insiders stealing data. He convinced the other Arbor executives to drop the contracting plans and bury his project.

One of @stake's young talents had worked out of the San Francisco office. Alex Stamos had joined not long out of UC Berkeley due to admiration for Mudge and the other founders. As @stake got subsumed by Symantec, he decided to start a new company with four friends. @stake had shown that it was possible to run a business that had a massive positive impact on the security of ordinary people. But it had two key flaws that he hoped to fix in the new company. The first was that it had taken venture money, which put it at the mercy of unrealistic financial goals. Declining outside investment money, Stamos and his partners, including Joel Wallenstrom and Jesse Burns from @stake, put up $2,000 each and bootstrapped the new consulting firm, iSec Partners. Instead of being heavy with management and salespeople, it operated like a law firm, with each partner handling his own client relationships.

The iSec model also attempted to deal with Stamos's other problem with @stake: that, in his words, "it had no moral center." Stamos made sure that neither he nor any of his partners would have to do anything that made them uncomfortable—any big decision would require unanimous agreement by the five.

iSec picked up consulting for Microsoft in 2004, after @stake was gone, and it helped with substantial improvements to security in Windows 7. Four years later, it got an invitation to help on a huge project for Google: the Android phone operating system. Android had been developed so secretly that Google's own excellent security people had been left out of the loop. iSec was called in just seven months before its launch. Among other things, iSec saw an enormous risk in Android's ecosystem. In a reasonable strategy for an underdog fighting against Apple's iPhone, Google planned to give away the software for free and let phone companies modify it as they saw fit. But iSec realized that Google had no way to insist that patches for the inevitable flaws would actually get shipped to and installed by consumers with any real speed.

iSec wrote a report on the danger and gave it to Andy Rubin, father of Android. "He ignored it," Stamos said, though Rubin later said he didn't recall the warning. More than a decade later, that is still Android's most dangerous flaw. Stamos was frustrated by being called in as an afterthought, and he began to think that working in-house was the way to go. Eventually, he joined internet mainstay Yahoo as chief information-security officer. Wallenstrom became CEO of secure messaging system Wickr; Jesse Burns stayed at iSec through its 2010 acquisition by NCC Group and in 2018 went to run Google's cloud security. Meanwhile, Dave Goldsmith in 2005 started iSec's East Coast rival Matasano Security, which attracted still more @stake alums to work from within to improve security at big software vendors and customers. He later became a senior executive at NCC.

The opening decade of the millennium was a strange and divisive time in security. "It was a time of moral reckoning. People realized the power that they had," Song said. Hundreds of focused tech experts with little socialization, let alone formal ethics training, were suddenly unleashed, with only a few groups and industry rock stars as potential role models and almost no open discussion of the right and wrong ways to behave. Most from @stake stayed in defensive security and hammered out different personal ethical codes in companies large and small. While they played an enormous role in improving security over the coming years, perhaps the most important work inspired by cDc didn't come from either corporations or government activity.

> # CHAPTER 9

> # TOR AND CITIZEN LAB

A T DEF CON back in 2001, as the Cult of the Dead Cow panel focused on hacktivism and touted the spin-off Hacktivismo, the group also announced what it said would be its first tool for evading government censorship. Dubbed "Peekabooty," the idea was complex. Users in free countries could install the software and then serve as intermediaries for people behind national firewalls in China or elsewhere, who might not be able to reach forbidden religious, news, or other websites directly. They could contact the volunteers running Peekabooty, who would not be blocked, and the volunteers could automatically route the desired content to them over the commonplace Secure Sockets Layer encryption, used at sites whose web address begins with "https." The authorities would not be able to read any traffic, and they would not be alarmed, because it would look like an ordinary encrypted business transaction.

Though the BBC had reported that the open-source project would be unleashed at Def Con, it was not ready for release. Laird Brown was hoping the advance publicity would attract more volunteers whom he could assign to different aspects of the job. The lead programmer to emerge and labor full-time on the task was software developer Paul Baranowski, who worked with Laird at Toronto start-up OpenCola. But Baranowski grew annoyed that Laird couldn't find other programmers, and he and friend Joey deVilla left Hacktivismo with the code. They released it on their own at a San Francisco conference in February 2002. "Hacktivismo is good with thinking up new projects" but not with follow-through, Baranowski said. But they didn't get critical mass, either. "Peekabooty's most valuable contribution was to say, 'Hey, this kind of thing is possible, here's an idea, go run with it," deVilla said. "Its truest value was as a proof of concept."

In 2004, Laird announced what he called the Six/Four System, a reference to the June 4, 1989, Tiananmen Square massacre. Written by incoming cDc member Kemal Akman, a talented German hacker with the handle Mixter, Six/Four was another serious try at a safe network of proxies. "I thought subverting totalitarian governments was cool," Kemal said. "cDc was making the most of its publicity for something positive." Kemal spent more than a year getting Six/Four to a point where it could be published for others to build on. But Hacktivismo's mailing list still only had about twenty active members and perhaps two hundred lurkers. Like Baranowski's earlier attempt, Six/Four failed to thrive in the wild. All the same, Hacktivismo's very public attempts to provide free secure web tools for the hundreds of millions of people under severe government scrutiny and restrictions inspired other programmers who got the job done. It turned out they didn't need to invent a new tool, just revise an old one.

Back in the mid-nineties, three men at the US Naval Research Laboratory had come up with the idea of bouncing internet traffic from one server to another to a third to keep both ends anonymous from snoops in the middle. The first node would know only where the initial contact had come from and, after opening the first layer of the message, where to send the rest of the content. The second would know only that it had heard from the first node and that the content needed to go to a third node. And the third node would know the final content and who outside the network should receive it. No one would know both the content and the sender. Because this multistep unpeeling resembled an onion, the project became known as the Onion Router, later abbreviated as Tor. The Defense Advanced Research Projects Agency (DARPA) provided new funding in 1997, seizing on the effort as a way to protect US military and other undercover officials from being identified as they investigated online.

To the government, though, this system had a fatal flaw: anyone who was contacted via Tor would know a fed was knocking on the door. But one of the original navy trio, mathematician Paul Syverson, along with new collaborators Roger Dingledine and Nick Mathewson, discovered a way to make it appealing enough that people outside the government would use it too, effectively hiding the agents in the crowd. They completed a prototype in September 2002, seven months after Peekabooty code came out, and released a version of Tor to the public the following year.

Peekabooty and Six/Four were major influences on Tor. "One of the strongest ways that Peekabooty influenced Tor was in pushing us to make good, clear specifications of how Tor works and what it tries to achieve," Dingledine said. In addition, he said, Peekabooty was years ahead of Tor in resisting censorship instead of just preserving anonymity. In 2004, craving funding from an outside and nongovernmental source, the Tor Project sought and won a grant from the Electronic Frontier Foundation, whose lawyers had already been involved in efforts by cDc and Hacktivismo. The EFF's support, in turn, helped Tor get money from Human Rights Watch, Google, and other parts of the federal government. Among other things, the early competition from Hacktivismo showed potential funders that there was a real demand for anonymity services and that activists independent of the government wanted to provide it. "We saw them as a key part of our constituency and fellow travelers," then EFF legal director and future executive director Cindy Cohn said of cDc. "These folks were trying to support the use of technology, sometimes really advanced stuff, to empower users and make social and political change. That's what we believed in too."

The friendly competition continued, to the good of the users. In 2006, Hacktivismo and a Texas cDc Ninja Strike Force member named Steve Topletz released the most popular of the group's anonymity tools, another try at a protected browser, called Xerobank, or xB. This one was designed to work with Tor, which at that point allowed for computer-to-computer connections, email, and other services but not easy web surfing. This browser was a modified version of Firefox that could work from a USB stick. That meant it could go with a user to a public computer and leave no trace. Once again, by publicly working on a safe browser, Hacktivismo prodded Tor along. Tor released its own browser as part of a bundle, making it far more usable. By 2006, more users were relying on Tor to evade censorship, not to stay anonymous, and China had become the third-largest market, with about ten thousand daily users.

In 2006, Laird organized a conference on wireless technology in Dharamsala, India, the seat of the exiled Tibetan government. That helped establish the area as a place for idealistic developers to work. Laird moved to the city in 2009 and spent three years there helping the community without pay. He worked on security in the Dalai Lama's office and helped build up local expertise. Then he spent two more years in Bangalore at an internet policy nonprofit.

Hacktivismo inspired hundreds or thousands of individuals and groups. Many had stories like Nathan Freitas's. A New York tech worker at the turn of the millennium, Freitas first heard about Tibetan repression from concerts headlined by the Beastie Boys. Through a work acquaintance, in the late 1990s he stumbled across a tiny Tibetan group in the Hell's Kitchen neighborhood that had only one modem and needed assistance setting up an office network. He did that, then noticed that there were viruses on virtually every machine. He realized that the Tibetans were under constant attack by the Chinese government.

In 2004, Freitas had to make a choice. The small start-up he helped found had been acquired by Palm, the smartphone pioneer, years before. Now Palm wanted to promote him and move him to Silicon Valley. But if he took that promotion, he'd be too busy for part-time activism. Freitas looked at what the hackers in cDc had been able to do. "They were hilarious, interesting, and effective," he said. They showed that small groups could "impact nation-state or global corporate policies. It was cDc that made me say, 'Maybe I can bring these things together.'"

Freitas quit Palm and used money from the acquisition to turn to hacktivism full-time. He went to China for a month with equipment to figure out how shortwave radio was being blocked and how to protect it. Then he helped start Tibet Action Institute with Students for a Free Tibet leader Lhadon Tethong, providing technical help and security advice to emigrants around the world. In the run-up to the Beijing Olympics in 2008, Freitas set up the satellite video feed to a publicly viewable website for a protest from the base camp on Mount Everest. The higher-profile activism brought more sophisticated cyberattacks from China, which just hardened his resolve. In just 2008, he equipped seventy people, many inside the mainland, with $3,000 crypto phones, burner phones, and netbooks. Freitas went to Dharamsala in 2008 to train Tibetans and met up with Laird. "He had this monk-like status, but he was this big, tall Canadian white guy," Freitas said. Laird coached Freitas and helped brainstorm about how to accomplish more with less, and he introduced Freitas to his world of contacts. When Google launched Android, Freitas jumped to use it for making a secure phone more cheaply. Eventually, he masterminded a version of Tor for the phones. Since then, his program has been downloaded 17 million times, and he now heads all of Tor's mobile offerings.

[x x]

Laird also inspired what many independent security experts consider the best model for researching and exposing government use of the internet for repression: the Citizen Lab, at the University of Toronto's Munk School of Global Affairs. It started with a University of Toronto student, Nart Villeneuve, all the way back in 2001. He had read cDc text files and was following along when the group launched Hacktivismo, soon joining that mailing list. Inspired, he launched a modest website tracking various hacktivism efforts, and he interviewed Laird for a text file of his own. "When I was starting out, I sort of became attracted to the mythical hacker archetype who could do everything," Villeneuve said. Without a technical background, he was interested in traditional politics and protests. Initially, disruptive tactics like web defacements and denial-of-service attacks made sense to him. But Laird's writing took him toward "a more constructive side of things," he said, including getting around censorship. At the time, people in China were complaining that they couldn't see some of the web, but there was no data about what was off-limits. Villeneuve came up with a way to test for website blocking and wrote a class paper on it for professor Ron Deibert. Deibert encouraged him to build out such software and hired him for what became the nonprofit OpenNet Initiative, which monitored censorship around the world. Then Villeneuve introduced Deibert to Laird.

The two men had long talks about the technological, social, political, and business challenges of keeping the internet as free as John Perry Barlow had declared it to be. They spoke about the need to get and publish objective, detailed information about what was happening inside routers and switches in hostile places. They agreed that the funding model for such a project had to be above reproach, so that it could not be accused of being in the pocket of an intelligence agency or a government. It would need to be able to get the word out to other researchers, the press, and the public, so that political pressure could be brought to bear on the implicated governments as well as companies, many of them based in the West, that provided the tools for censorship and spying.

"Some of our early interactions around hacktivism definitely were important to me in terms of setting up Citizen Lab," Deibert said. "I was, like Laird, inspired by this hacking in the original sense of the word, combined with some political orientation or morality underneath it. I thought that was very appealing. I think we have the same outlook and philosophy about what's acceptable and not."

Major research belonged at a university, Laird argued, because it prioritized scholarship over profit or politics. Like Tor in the practical sphere, a university could take in some government money yet remain clean, as long as it declared its policies. It could also draw on those in multiple disciplines—computer and security experts but also political scientists. It would take a tremendous leap in ambition, since many colleges still didn't even offer courses in security.

In the spring of 2001, after receiving approval from the University of Toronto and an initial grant from the Ford Foundation, Deibert opened the Citizen Lab, with Villeneuve as his first hire. The modest official mission: to study cyberspace "in the context of international security." But the tools to be used ranged from technical exploration to field research to political theory. Almost immediately, the September 11 attacks multiplied the stakes. With US intelligence agencies lambasted for not knowing enough, surveillance was bound to soar in the West as well as the East. And that was just the beginning. The geopolitics of the internet were metastasizing, on their way to becoming one of the most significant and complex issues facing the world. It would be hard to find the answers. But no one would be in a better position to try.

Early on, the lab looked hard at web filters in the Arab world, including their suppliers and what pages or words were restricted. As part of that long-running effort, it found that Syria was using software from Silicon Valley firm Blue Coat to spy on its people, potentially violating US sanctions. The lab also took on the legal sale of exploits and other tools for what the industry calls "lawful interception," tracking many cases where the vendors said they sold only to governments that respected human rights. Despite such claims, researchers often found repressive regimes deploying wares from companies like UK- and Germany-based Gamma Group and Italian firm Hacking Team against human rights advocates, journalists, and minority-party politicians. Much later, a devastating series of four front-page reports in the *New York Times*, driven by Citizen Lab findings, documented Israeli company NSO Group's Pegasus spyware being used against Mexican journalists, politicians, and others in Mexico, including officials investigating mass disappearances and even anti-obesity campaigners. Mexico's president ordered an investigation that the FBI concluded was a sham.

Time and time again, the lab's independent academic structure gave it a way to write about what others could not. The university's review board had to approve research methods on ethical as well as legal grounds. All the same, the retired head of Canada's main intelligence agency once noted pointedly that some people thought Deibert should be arrested. As more countries turned to spying on each other over the net, using companies as stepping-stones or knowing accomplices, untangling it all could have had political and business repercussions for any private researchers. The same big companies that excelled at examining and explaining malicious software that served organized crime shied away from being as clear when they realized that the culprits were the governments controlling major markets for their security software. Governments themselves stayed mum because the intelligence agencies maintained dominance over cyber offense and defense within the bureaucracy, and such agencies preferred not to reveal what they knew.

Some specialized firms, such as Mandiant and CrowdStrike, disclosed more in private reports to clients, and they sometimes went public with accounts attributing infections in certain industries to coordinated campaigns by government-affiliated hacking groups. But they faced accusations of bias because their detection systems were only deployed in some countries, they had US government contracts, or they had marketing reasons for publishing what they did. Moscow-based Kaspersky Lab, likewise, became the best in the world at ferreting out US-sponsored cyberespionage campaigns, beginning with Stuxnet, the pathbreaking weapon that defanged Iranian nuclear centrifuges before its exposure in 2010 opened everyone's eyes to the new era of cyberwarfare. But Kaspersky found very little new to say about Russian malware.

Citizen Lab could call things as it saw them. And it extended its reach by working with researchers inside other companies, including Google, who would have found it hard to publish under the name of their principal employer. The lab also worked with researchers at Amnesty International and the Electronic Frontier Foundation.

[x x]

The lab's work only got better and more important as the years went on. One of Villeneuve's biggest projects was learning what spying was happening in Tibet. There was likely to be at least some, since activists were routinely stopped from entering China, at times arrested, and occasionally shown transcripts of their electronic chats with people inside China. They were risking their lives. Laird introduced the team to Greg Walton, who was still spending time in Dharamsala and working on Canadian-funded rights initiatives. Walton had good relations with the Tibetans, and Deibert hired him as a field researcher in 2008. For the first time, Deibert learned about targeted malware. Later, Walton got the Dalai Lama to agree to turn over the leadership's computers for study. Attackers had riddled those machines with compromises. But a hunch led to a big payoff. The network traffic from many of the machines included the same string of twenty-two characters. So Villeneuve googled that string. In short order, he was on a machine in mainland China, looking at a portal listing hundreds of computers that same group had broken into. The victims included an email server for the Associated Press in Hong Kong, an unclassified computer at NATO headquarters, and embassies belonging to India, Pakistan, Germany, and Thailand.

Deibert's team dubbed the spy network GhostNet. Citizen Lab disclosed it in 2009, making front pages around the world. Deibert had brought the *New York Times* in early, in part for maximum impact and in part as a hedge in case the Canadian government tried to suppress what the Citizen Lab had found. The first such account by a nongovernment agency and one of the first of any kind linking specific computer espionage to a world power, the GhostNet paper did not explicitly blame China. But that country was obviously behind this instance of what would become known as an advanced persistent threat, or a committed cyberspace adversary. Four servers controlled the penetrations, including one on the island of Hainan, home to the Third Technical Department of the Chinese People's Liberation Army.

The team had worked feverishly to unwind all of the connections and document how they functioned. At the same time, the crew had wrestled with new issues about disclosure. If China had been found spying on a single identifiable person, they would have felt a responsibility to warn the victim, though there was no clear ethical rule on the matter. What if their own government was among the victims? What about other governments? Who should be told what, and when? Rather than go directly to Canadian intelligence and risk being co-opted, Deibert went to the Canadian computer emergency response team as a courtesy. The Citizen Lab also asked the Canadian foreign affairs ministry if it could pass along notifications to other countries. It took months for the ministry to even respond, and then it declined to help.

Nathan Freitas, the hacktivist and Tor specialist helping Tibetans, had run into a similar problem. As reports shed light on Chinese spying, more people came looking for copies of what infected the Tibetans. "Malware we got hit with, no one had ever seen before," Freitas said. "Researchers came out of the woodwork saying, 'Can we have a sample?'" Some were academics looking for material for doctoral theses, some were employees of private companies, some were government officials. It was obvious that some were intelligence agents or contractors. "You can't fool yourself," Freitas realized. "This is global cybersecurity warfare." Rather than try to sort out who was working for whom and whether it would be appropriate to favor one country's emissaries over another's, Freitas threw up his hands. He decided to share samples only with the Citizen Lab, which had done the hard ethical reasoning. But in a community as heavily targeted as the Tibetans, with various Westerners helping out, some information inevitably went to the powerful Western agencies who were fighting the Chinese in many places. Hacktivism gave those operatives an excuse to be in and among the activists.

For all of Deibert's careful ethical balancing, intelligence figures still involved themselves in the Citizen Lab's work. The lab got analytical help from Rafal Rohozinski, a lab research advisor who was wearing other hats at the same time. Rohozinski was CEO of Psiphon Inc., a proxy network for evading censorship that the Citizen Lab had spun out. He also had worked in the military and as a technical advisor to the UN on telecommunications projects in former Soviet countries around the world. Though he described himself as an independent contractor, he acknowledged an intelligence background, and his affinities were clear. Laird and Villeneuve both called him a "spook," which Rohozinski said was inaccurate.

Laird also denies being a spy, and he never revealed himself as one to cDc. But his odd initial approach to the group, changing Hong Kong Blondes backstory, and later international work have caused several in cDc to wonder, even without being aware of the intelligence relationships that have since come to light. Laird was close enough to the community that it may have regarded him, fairly or not, as an "asset," two of his intelligence contacts told me. That's enough to change the history of hacktivism.

Intelligence agencies ardently consumed information from the GhostNet effort. Rohozinski and Villeneuve briefed the NSA together, and Rohozinski learned more from Greg Walton directly. In some ways, Western intelligence agencies delighted at the Citizen Lab's work. It exposed a geopolitical rival, and it looked better because the lab had no ulterior motive. It also engaged in legal but invasive use of internet tools, such as port scanners, that would have required multiple levels of approval if some governments had used them directly. Yet Deibert detected hostility as well from the Canadian authorities he ran into, more than he could explain as professional jealousy or disdain for upstarts. Poring over the documents released by Edward Snowden a few years later, Deibert thought he realized why, and Rohozinski agreed: the Canadians had known about the Chinese spy network and had been piggybacking on it, collecting their own intelligence, until the Citizen Lab blew the whistle.

The year after the GhostNet report, Google said that the Chinese had hacked it as well, and that it was pulling out of the mainland as a result. Now everyone realized that they had been living in an undeclared cyberwar. Google had among the best technical defenders anywhere. After Google realized the Chinese had gotten in and gone after the accounts of human rights advocates and Google's own code, it brought in the best outside minds it could find. That included Dave Aitel and other NSA veterans, and even the NSA itself. The public was alarmed, but it never realized how effective the Chinese campaign really was, because no one had an incentive to admit it. According to Mudge, the Chinese had broken into repositories for the source code of many big companies and written in what looked like programming mistakes. In reality, they were back doors that would allow Chinese spies to break into the customers of those big tech companies whenever they wanted. In a fight like that, Google and many others understandably considered the NSA to be the good guys. But it was not that simple. In a few years, with the public debut of NSA leaker Edward Snowden, Google and many other American tech companies, to say nothing of the rest of the world, would see the agency as an archenemy.

> # CHAPTER 10

> # JAKE

AFTER HACKTIVISMO'S XEROBANK browser helped drive the Tor
Project to broaden its mission in 2006, the service became truly
useful to large numbers of people. Tor began spreading in earnest in
countries like China and Iran, where surveillance could be swiftly
followed by jail time. Psiphon, Freegate, and other services could deliver forbidden parts of the web to readers, but only the souped-up Tor
could both serve up such destinations and obscure who was reading
them. Not coincidentally, US government funding for the Tor Project
increased substantially that year. As with other free-communication
projects, the greater the take-up in areas ruled by figures both opposed
to American interests and repressive to their own locals, the greater the
US enthusiasm for tools boosting free speech.

But Tor's origins inside a navy lab and its ongoing federal funding gave room for suspicions about whether it contained a hidden back
door for US spies or was otherwise corrupted, even though its source
code was public for review. It was not, as Edward Snowden's documents
would show years later. Tor frustrated US intelligence agencies, which
were unable to crack it reliably. Support from the Electronic Frontier
Foundation and endorsements from public-minded cryptographers, including some on the Cypherpunks mailing list, helped convince many
that they could trust Tor. But a majority of that crowd were of a previous generation, long-haired mathematicians more comfortable in a

university library or the bowels of a Silicon Valley office park than hanging out with young activists.

The apparent answer to Tor's public-relations problem arrived in the person of Jacob Appelbaum, known in person as Jake and on Twitter as @IOerror, a reference to a malfunction in input/output processing. Jake was young and good-looking, an engaging public speaker and a frequent presenter at serious security conferences. He also had an extraordinarily compelling personal story. If many hackers turned to computers early to escape hard childhoods, Jake's case was extreme. His mother, a schizophrenic, raised him until she lost custody to an aunt, who left Jake in a group home. He went to his father at age ten, but the man grew addicted to heroin. Father and son lived on buses and in drug dens, and Jake once found his dad overdosing and near death. Returning to group homes, Jake dropped out of high school and taught himself to code, working for the likes of Greenpeace and the Rainforest Action Network. Jake met Tor leaders Roger Dingledine and Nick Mathewson at a Def Con and began volunteering. He joined as staff in 2008 and quickly became Tor's most visible spokesman. He was also among the best traveled in Tor's network, flying to hotspots around the world to teach locals how to use it.

Wherever the attention was, it seemed Jake was there too, even as coauthor of a research paper showing that one could recover plaintext passwords by suddenly freezing a computer's RAM data storage. "Pretty neat," Luke Benfey wrote when he successfully nominated Jake for Cult of the Dead Cow membership in 2008. "He is certainly enthusiastic," Luke added, though "a little bit weird." Most of the core cDc members at that point were impressed enough to support the motion, and Jake was in with Kevin Wheeler's final blessing. Even those who had not met him felt like they knew him because his story had been told by security, tech, and even some mainstream press outlets. There was an added attraction because the ranks of the group were thinning, and recruits with younger followers had to be prized if cDc, already more than twenty years old, could continue as a vital entity.

Laird Brown had brought in Kemal Akman, known as Mixter, and others through Hacktivismo, and old friends like Patrick "Lord Digital" Kroupa had joined. Some fresh security researchers like Adam O'Donnell also came aboard. But more were asking to be taken off the internal mailing list. That included both some of the technological powerhouses, busy running their own companies, and old-timers who were less technical, like Carrie Campbell. In 2006, she wrote with sadness and asked to go off the list, partly blaming herself for failing to get to know new members and drifting away.

> I'm afraid my interests in the hacker scene have waned long ago. You new people don't know me. I was a 16 year old girl when "Psychedelic Warlord" saw my crazy, poorly-written teenage angst postings on his BBS and invited me to join cDc. I joined happily, honored, and proceeded to write crappy, horrific, 16 year old bloody t-files. I loved the community of smart people (and their girlfriends) to converse with and bounce ideas off of. The acceptance of my female gender is extremely rare in the hacker scene and I appreciate it. I never pretended to be a hacker, since I'm not skillful in that area (though social engineering came easily to me).
>
> Somehow I ended up purely by accident as the only girl in the world's most notorious hacker group, and while that was enormously amusing, I am now approaching 40 years old rapidly. I have no energy left for cDc or the mailing list. I do have energy for the wonderful friends I made throughout this oh gosh, 21 or so year journey. Please do email me from time to time.

Because she had been a core human connector within cDc and went back nearly as far as he did, Carrie's departure moved Kevin and made him worry that others would follow her. He took a long walk through Central Park, then wrote and asked the others to stay. "'The hacker scene' isn't something I'm into, other than as a recruiting pool for sharp motherfuckers—and a hearty 'hell-yeah' for being that," Kevin wrote.

"Someday, hopefully there'll be AIs in our mix and we'll be trying to make practical sense of geopolitics and philanthropy. There's always more to say, more to point out, more that's fascinating and awesome and exciting. The universal, rock-solid, eternal part is the sharing, the communicating. Anyways—I want you guys to stick around."

[x x]

But the herd needed new blood. If Jake was as good as he seemed, he could bring not only new energy but potentially more recruits. Jake soon gave some evidence of being a good bet. His press clippings were astonishing, including a 2010 *Rolling Stone* profile that called him "a bizarro version of Mark Zuckerberg" and the leading spreader of "the gospel of anonymity."

Inside cDc, Jake handled himself differently than the others, arguing more fiercely and sometimes with disdain for his elders. That accelerated after he hooked up with something even bigger than Tor: WikiLeaks. Activist hackers started the site in 2006 and first won wide attention in 2010, when they posted a video called "Collateral Murder" that captured the gunfire from a US helicopter that killed a dozen people, including two Reuters journalists, in Iraq. The video disproved US claims that the shooting was part of a battle.

The one WikiLeaks founder who would be left standing after years of internal dissension and splits was Australian Julian Assange, who had nearly as bad a childhood as Jake, including hiding with his mother from a vengeful cult. Even more of a show-off than Jake, Assange had been a belligerently antiestablishment and sometimes malicious hacker in his native Australia. Under the name "proff," he had been on some of the most popular Internet Relay Chat channels devoted to security and hacking in the 1990s, including #hack. He was an ambitious and dangerous hacker, later claiming credit for breaking into Australian government computers and backdooring the Pentagon. He was not remembered fondly by cDc, which saw him as an egotist who usually lurked instead of contributing to discussions. When he did speak up, it was often to criticize or ask for working code he could use to break into targets.

Assange was also a regular presence on the Cypherpunks mailing list in 1996 and 1997, comparing notes with others about developments in cryptography and ongoing tensions with officials of many governments bent on restricting it. Assange advertised his own mailing list for "legal aspects of computer crime" as well, opening with a manifesto that declared computer crimes were being overprosecuted and that intrusions should not be considered criminal acts unless they caused harm. At one point he posted about a commercial spam operation and asked: "Who wants to take this site down first?" Assange and Mudge treated each other with respect, however, and met for dinner at the Chaos Computer Club's 2009 gathering in Germany before they aligned with opposite world powers.

cDc admired much about the early WikiLeaks, with good reason. The site published a wide variety of documents and seemed most focused on government wrongdoing. When it obtained tens of thousands of US State Department cables from then Private Bradley Manning (now Chelsea Manning) in 2010, it worked with media partners that sifted through for important stories while not printing information that could lead to the deaths of those cooperating with American officials abroad. "I have quite a few issues with the organization, but I like it more than I dislike it, at least for the time being," Laird wrote to the cDc list that year.

Assange was to speak at the HOPE conference in New York in July 2010. But the Pentagon had labeled WikiLeaks as a threat, and Assange feared arrest. Jake appeared by surprise instead. He gave a fiery recounting of the whistle-blowing site's history and courage, which he said continued the tradition of the *Washington Post* and the *New York Times* of Watergate and Vietnam coverage, before more recent timidity such as the *Times*'s yearlong delay in exposing warrantless wiretapping by the NSA. "When the media is gagged, we refuse to be gagged,"

Jake said. He added that he wouldn't say anything about hacker Adrian Lamo, who had turned Manning in to the authorities after the troubled private confided in him that he had leaked State Department cables. Then Jake unbuttoned his shirt and revealed a T-shirt underneath that read: "Stop Snitching." At the end of his talk, the room suddenly plunged into darkness, and the lights came up to show what appeared to be Jake being ushered out to safety. In reality, he was a body double, deployed in order to stop Jake from getting arrested or hurt, or simply to convince the audience that either was a possibility. The actual Jake had gone out the back.

After that, American customs and border patrol officers often stopped Jake at airports and interrogated him without charges. He complained vociferously in public and to his fellow cDc members, telling them in early 2011 that "the U.S. government has flagged me just as the Nazis forced Jews to wear a gold star. I don't have the choice of removing my marks, though, they're in the passport system for life." As someone who lived on the internet and credited it with saving him as a kid, Jake would have been aware of Godwin's law. Named for its originator and EFF's first staff attorney, Mike Godwin, the aphorism states: "As an online discussion grows longer, the probability of a comparison involving Nazis or Hitler approaches 1." Godwin was mourning both the declining quality of online discussion and the lack of gravity owed to the Holocaust.

The old-timers in cDc were not impressed. "Dude, seriously?" Luke wrote. "You just managed to pull off the elusive one-man Godwin. Jake, I think you need to have some understanding that you've made this bed, and now you have to lie in it." Prosecutor Glenn Kurtzrock was more precise in referring to the rules governing US Customs and Border Protection. "It doesn't appear that CBP did anything wrong. They are entirely authorized to search and detain you when entering the country under the U.S. code, including the contents of any electronic devices." Jake also sparred with the others repeatedly over Assange, whom Laird said was about as democratic in management style as the ruler of Saudi Arabia. "So much for hacktivist solidarity," Jake complained. Luke and

Kemal took a middle ground: Assange was an asshole, but he seemed to be doing good things.

Broadly speaking, the State Department cables released by WikiLeaks showed US officials doing their jobs. There was no great sinister conspiracy. But the various stories still embarrassed the American government and hurt diplomatic relations. The cables contained candid assessments of foreign heads of state, including their unsavory alliances and appetites for corruption. The antisecrecy fervor at WikiLeaks stoked a rollicking debate inside cDc. Glenn and others saw Assange as reckless, noting that the judicial system and other parts of government have very good reasons for keeping some facts confidential. Arguing out a hypothetical about missile launch codes getting into the wrong hands, Jake declared: "Perhaps you shouldn't have missiles to launch if you can't keep your codes secret?" Jake said a lot of provocative things, declaring that wiretaps were "entirely bogus" and that most search warrants were improper. One of the most surprising assertions came in response to questions about who should decide what secrets to publish. Instead of WikiLeaks holding that right as a publisher, Jake said it was up to WikiLeaks's sources, whoever they were. "It's a rough reality, but bitching about WikiLeaks makes little to no sense," he wrote. "The point of the press is to inform."

[x x]

Members of Congress condemned WikiLeaks, and a federal criminal investigation put pressure on PayPal, Visa, and others that helped people donate to the website. The sprawling online activist group known as Anonymous then coordinated denial-of-service attacks on PayPal and Visa, effectively commandeering the mantle of hacktivism. The story of Anonymous, told more fully in books by anthropologist Gabriella Coleman and journalist Parmy Olson, is fascinating and complex. It also owes a little of its culture to cDc. One of cDc's good friends and onetime web hoster, Tom Dell, had written software for Patrick Kroupa's Mind-Vox and then run Rotten.com, an early shock site that was a forerunner of 4chan. 4chan was mostly teenage boys chatting about pictures, and posts were labeled "Anonymous" by default. But it had flashes of political action when core internet values, such as freedom of speech, were threatened. When the Church of Scientology tried to suppress publication of its secrets, 4chan users coordinated online and real-world protests, and the participants spun off as Anonymous. Subsequent targets included copyright enforcers such as the Motion Picture Association of America. From the beginning, corralling massive crowds in Internet Relay Chat into something productive was extremely difficult. Organizers would peel off into secret smaller channels to thrash through priorities and then return to the larger gatherings to spread the word.

Anyone could declare themselves a member of Anonymous, and any member could call for an operation, most commonly a denial-of-service attack. It was up to other members whether to participate in any of the operations. With the denial-of-service attacks, members were encouraged to download a tool that would let them participate. But while that let participants feel like they played an important role with little risk, neither conclusion was justified. Some were arrested, because the tool did not hide their IP addresses. And most of the real firepower came from botnets, networks of captured machines controlled by a small subset of Anonymous members. The regular members were helping to provide cover and confusion, and that was about it.

As Anonymous allied with WikiLeaks and struck the payment sites with denial-of-service attacks, cDc members split on the ethics of the issue and opted to do nothing as a collective. Laird, who had been giving speeches for years on the ethics of hacktivism, carried the most weight on the subject. He opposed the denial-of-service attacks as censorship, arguing that the cure for bad speech is more speech. As reporters sought him out for comments about Anonymous, he stood firm. Luke, on the other hand, held that some denial-of-service attacks were reasonable civil disobedience, depending on the motives and targets. The onslaught only temporarily disabled PayPal and Visa while they shored up their defenses, he said. But knocking them briefly offline brought media attention and greater awareness of the issues involved.

When the focus of crowds is one of the few things that can change policies, Luke felt, it made for a decent trade-off.

Dozens of Anonymous members did have hacking skills, as became clear after I wrote a short 2011 story in the *Financial Times* about a researcher, Aaron Barr, who said he would give a conference talk about the people he believed led the group. Highly skilled Anonymous ringleaders had a private channel for communication, and after my story appeared, the members of that channel broke into the files of Barr and two affiliated companies, HBGary Federal and HBGary, in part to make sure he didn't have the goods on them. They published emails from the companies that showed that Barr was off the mark and that he was engaged in some questionable pursuits, including seeking a deal to discredit WikiLeaks by supplying faked information.

The ace hackers announced themselves to the world as Lulz Security, began tweeting as @LulzSec, and went on a wild performance-art run, hacking Rupert Murdoch's tabloids to post stories announcing his death and even taking requests from their followers. LulzSec kept up a prolific and funny Twitter stream, largely manned by Topiary, later identified as eighteen-year-old Shetland Islands resident Jake Davis, and updated a web page with a logo and the slogan "Set sail for fail!" In an anonymous interview shortly before his arrest, Davis explained why he thought LulzSec had so much of the public behind it: "What we did was different from other hacking groups. We had an active Twitter (controlled by me), cute cats in deface messages, and a generally playful, cartoonlike aura to our operations. We knew when to start, we knew when to stop, and most of all we knew how to have fun."

Davis later said he had been inspired by UK satirist Chris Morris and comedian Noel Fielding, and that his playfulness had a serious point: he wanted people to wonder why major security failures were so common, instead of attributing all breaches to unstoppable geniuses. "It was a mix of deliberate absurdity [and] a carefree childishness that was intended to alter the conversation to 'These people are clearly just doing this as a game. Perhaps we should actually start thinking about security if these morons can wreak this much havoc.'"

The stunts and public commentary echoed the Back Orifice performances. Davis had honed his writing by drafting entries for the satirical, inside-hackerdom site Encyclopedia Dramatica, which looked a bit like old-time cDc text files. In person, Davis was quiet and shy—quite like cDc founder Kevin Wheeler offstage. But the serious illegal acts put Lulz Security on a different path, and in any case it would have lacked the stability of the Cult of the Dead Cow. That's because the members did not know each other in the physical world, so they could not make good decisions about trust. That problem was multiplied a thousandfold in Anonymous writ large. All the same, Anonymous and LulzSec launched a new era of stealing and publicizing material in a manner that was claimed to be for the public good.

Many of the LulzSec capers were driven by both politics and entertainment value. Toward the end, after puzzling as cDc did about what to do with all the attention, Davis announced that LulzSec would revive Antisec, an old campaign against white-hat security professionals. This time, LulzSec would ally with the broader Anonymous and go after government security agencies, banks, and other establishment powers. Julian Assange was tracking events closely, at one point contacting the group for help getting into Icelandic email services that might show that government treating WikiLeaks unfairly. After LulzSec supporter Jeremy Hammond hacked US intelligence consulting firm Stratfor, WikiLeaks published millions of Stratfor emails with clients. Eventually authorities caught almost the entire LulzSec crew. Technological ringleader Hector Monsegur, alias Sabu, flipped and helped put Davis and the others away. After he began working undercover for the FBI in return for a radically reduced sentence, Monsegur encouraged hackers to disrupt more targets, and he repeatedly reached out to Assange and Jake, which suggests both were under US investigation.

The FBI was not the only agency to infiltrate Anonymous. Taking advantage of its loose structure, ordinary criminals used a group protest of Sony Corporation policies to break in and steal credit card numbers. Russia also had a substantial presence in Anonymous. In retrospect, it is interesting that some Anonymous members would later go on Moscow's payroll. One of them, Cassandra Fairbanks, moved from real-world Anonymous demonstrations, to attending and writing about Black Lives Matter protests, to avidly supporting Bernie Sanders in the 2016 primaries. With more than a hundred thousand Twitter followers, she then took a job at the Russian propaganda outlet Sputnik and switched to full-throated support for Trump through the 2016 general election and afterward. Just before the November vote, she appeared on Alex Jones's YouTube conspiracy channel, saying it was "pretty likely" that emails hacked from Hillary Clinton campaign chair John Podesta's Gmail account contained coded references to pedophilia.

Monsegur liked to talk about his political work. He told journalists that he had hacked for a cause long before, protesting US Navy test-bombing in Puerto Rico, where his family had lived. He also claimed to have defaced Chinese websites in 2001, as other Hacktivismo supporters did. Monsegur said he joined Anonymous as it fought PayPal and Visa and moved up from the cacophony of the main Internet Relay Chat channel to more elite planning channels, including the one that morphed into LulzSec. The most impressive story: as part of Anonymous's Operation Tunisia, during the Arab Spring democratic uprisings, he personally defaced the web page of the country's prime minister, who had approved mass hacking of citizens. But that and the other relatively high-minded feats proved impossible to confirm. Author Olson described the Tunisian defacement as Monsegur's work, citing him as the only source. Professor Gabriella Coleman, who was perceived as sympathetic, obtained chat logs and said Monsegur did not lead the team that performed the Tunisian defacing. In any case, even Monsegur's few remaining supporters would have to agree he was an inveterate liar. His more prosaic crimes, such as stealing car parts and credit card numbers, were no mystery at all.

Another core LulzSec member, sixteen-year-old Mustafa "tflow" Al-Bassam, an Iraqi refugee in London, did something more challenging than defacing a website. With help from a local Tunisian who got trick phishing emails from the government, Al-Bassam hacked into the server sending the emails and modified the malicious program they carried, quietly rendering it impotent.

Like Monsegur's, Assange's judgment was soon called into doubt. Wanted for questioning in a Swedish probe of sexual misconduct, Assange lost a bid to avoid extradition and jumped bail in 2012, fleeing into Ecuador's embassy in London and remaining there. After Assange railed against his Swedish accusers from hiding, some of those inside cDc who had reserved judgment about him moved into the opposition. But as that furor grew and WikiLeaks increasingly focused on exposing US secrets, Jake stayed the course. That loyalty built his stature as an information-security rock star for those who remained believers in Assange. Within cDc, however, he caused more friction.

Laird wrote to the private cDc email list that he was concerned about the departures of other WikiLeaks stalwarts fed up with Assange's dictatorial ways and grandstanding. That meant that the group depended on one man, who was showing himself to be less and less dependable. "I had heard that Assange had problems with women months before any of this Swedish thing became public," Laird wrote. "Does Assange tone down his profile until the rape cloud is lifted, Hell no. He can't be in front of the press enough. So if he's convicted of some sort of sexual misdemeanor this will—in my opinion—completely torpedo WikiLeaks." Jake came up firing, defending Assange as a visionary and dismissing the female complainants as "fame seeking."

[x x]

WikiLeaks's flagging reputation was one reason Edward Snowden did not turn to it with his documents in 2013, though Assange did later dispatch a colleague to spirit him from Hong Kong to Moscow and asylum. Inspired by John Perry Barlow's independence declaration, Snowden wore an Electronic Frontier Foundation sweatshirt on the job at the NSA. When he felt compelled to warn the world about what his agency had been doing, Snowden first reached out anonymously to a new EFF spin-off called the Freedom of the Press Foundation, which had been formed in support of WikiLeaks by Barlow, Pentagon Papers leaker Daniel Ellsberg, *Boing Boing*'s Xeni Jardin, and a few EFF staffers. One of the staffers recommended Snowden get in touch with Freedom of the Press Foundation director Laura Poitras, who had been making a movie about WikiLeaks, and former Salon columnist Glenn Greenwald at the UK's *Guardian*. The *Guardian* published many of the most important revelations from Snowden's trove, but the pair also collaborated with other publications, including the *Washington Post* and the *New York Times*, to write up Snowden's disclosures.

Jake later reported related stories for *Der Spiegel* in Germany, going further in exposing specific US capabilities instead of broad practices. Though it was widely assumed the documents referred to in the stories came from Snowden, the information they contained has not been cited by the *Guardian*, *New York Times*, or *Washington Post*, which all had access to the main Snowden archives. That suggests a few possibilities: *Der Spiegel* may have had a different standard about what to publish, the material may have come from a second, still-unknown source, or it may even have been obtained through hacks by the Russian government, which then leaked to *Der Spiegel*.

Snowden showed how closely the US government worked with and through American technology companies, sucking up domestic calling records, sifting through emails for specified content, and examining communications in other countries, which are not protected by the Constitution's prohibition on unreasonable searches and seizures. Google, for one, had not realized that the NSA was breaking into its properties overseas, and it moved swiftly to encrypt internal transfers of user data. Other stories showed that the NSA had continued to corrupt security products by paying for back doors to be implanted or by promoting standards that it knew it could break, such as the Dual Elliptic Curve pseudo-random number generator. No major reforms passed Congress, and the anger in other countries hastened the balkanization of the internet and sped up the introduction of nationalist technology policies that hurt US providers, to the detriment of populations everywhere. At the same time, the revelations intensified work on more secure alternatives.

One of the most promising was Signal, developed by a team led by the brilliant anarchist and ex-hobo known as Moxie Marlinspike, and released in 2014. The Snowden disclosures carried enough force that Signal's end-to-end protocol became mainstream even without most of its users' knowledge. The two founders of WhatsApp, an enormously popular messaging app for smartphones, were Jan Koum and Brian Acton. They sold the company to Facebook in early 2014 for $19 billion and stayed to run it with some independence. Koum belonged to the long-running hacking group w00w00, which included cDc's Adam O'Donnell and such cDc friends as Dug Song. Song urged Koum to get in touch with Marlinspike, and Koum agreed when Acton proposed having WhatsApp adopt the nonprofit Signal's open-source technology, protecting a billion people from mass surveillance. In 2018, Acton would donate $50 million to create a new foundation to spread Signal much further and sign on as executive chairman, citing the opportunity to "make a meaningful contribution to society by building sustainable technology that respects users and does not rely on the commoditization of personal data." Later, he said he had been motivated "by an increase in requests from law enforcement and the desire to render those requests useless." Koum stayed on at Facebook, where he was one of only three executives also serving on the company's board. Though he continued to run WhatsApp, Facebook began demanding more data than expected about WhatsApp users, building up ad revenue but also exposing the users to greater corporate and government scrutiny. Koum would quit in mid-2018.

[x x]

Jake moved to Germany in 2012 and spent more time promoting Tor than he did coding for it. He attached his name to security research on other issues that drew wide attention, but some coauthors later complained that he had asked to be added so that he could use his fame to promote the work.

Jake flouted his edginess in multiple ways, including boasting of his past work for San Francisco bondage porn site Kink.com and sexually propositioning people at first meeting, even in professional contexts. He bragged of multiple lovers and had relationships with filmmaker Laura Poitras, who later acknowledged that he had mistreated a friend of hers, as well as *Boing Boing*'s Xeni Jardin, a friend to several in cDc. Jake spoke of waking up in bed with Assange and two women, and he attended private sex parties (less rare in hacker culture than elsewhere). Even there, he pushed past the norms of the environment.

One of his techniques in pursuing sex from someone who might otherwise object was to begin transgressive behavior in front of another senior hacker, said longtime friend Andy Isaacson. That hacker, not wanting to burn a relationship, would not object. This in turn put more pressure on the prey, who was more likely to assume that Jake was following norms in the situation or would have a witness on his side if not. "Jake's magnificent gifts overlap with the same fundamentals as his failure. He's very intelligent, and he doesn't let things go," said Isaacson. A key lesson from the experience, he said, is that "abusers can use loose organizations as hunting grounds."

As a champion social engineer, Jake exploited his role as a gateway to hacker prominence, victims said, leading many to conclude they would be frozen out if they objected. He targeted more junior people in the Tor community, where complaints led to a ten-day suspension for suspected harassment in the spring of 2015. That did not dissuade him. Fortunately, longtime EFF head Shari Steele took over as Tor executive director later that year, bringing more responsive leadership.

Steele came too late for some, including a young engineer named Chelsea Komlo, who had gravitated toward security after hearing Jake speak at her company about Snowden's leaks. Komlo traveled to Hamburg for the Chaos Computer Congress in December 2015 and went to Berlin with others after to socialize. At Jake's apartment the night of

January 1, she blacked out and woke to realize Jake was having sex with her without consent. Earlier, she had refused his repeated requests to have sex in front of and with others, but both of those things occurred. Back home and upset, she confided in people who knew other victims, and she got in touch with them. Steele's arrival at Tor gave them hope that change was possible. To protect themselves and warn others, they went to Steele and also prepared a website where they told their stories of assault and coercion under pseudonyms. "For me, it was really important that new people entering the community not have what happened to me happen to them," Komlo said.

Jake resigned on Thursday, June 2, 2016, but Tor gave no reason in its announcement. Only after the anonymous website went live the next day did Steele acknowledge, on Saturday, that concrete sexual assault allegations and an investigation were behind Jake's departure. At various times over the next year and a half, some victims identified themselves, including Komlo and Leigh Honeywell, a Canadian security engineer for big tech companies. Honeywell said that during an on-and-off consensual relationship a decade earlier, Jake had ignored a safe word and become violent. "Being involved with him was a steady stream of humiliations small and large," Honeywell wrote on her own site. "He mistreated me in front of others and over-shared about our intimate interactions with friends who were often also professional colleagues."

Without criminal charges, Jake fought back, in part through media connections who cast doubt on some of the anecdotes. He denied the worst accusations, threatened legal action against the women, and implied that the attacks against him stemmed from his work for free speech and secure technology. Still more people came forward, and the weight of evidence against him grew. "Tor handled it in a way that you would hope and expect," Komlo said. Komlo was invited to a Tor conference the next year, began writing code for the project, and later was designated a core contributor. That was especially encouraging, Komlo said, because of the male dominance in the field and because women are more likely to be abused by men who spy on them. "Security and

privacy is a great field for women, because there is a lot of moral reasoning, and you are in it because you want to protect people, and that
should be something that resonates with not just straight white men."

The Tor Project replaced its entire board. Even Jake's mentor,
Roger Dingledine, and Nick Mathewson stepped down while remaining
lead employees. People involved in the process said that the prior regime had had a leadership vacuum and consistently played down what
many people told them about Jake. "What you tolerate and don't tolerate defines you," one of them said. New directors included the EFF's
Cindy Cohn, cryptography experts Bruce Schneier and Matt Blaze, and
Gabriella Coleman, the anthropologist who chronicled Anonymous. After a few days, Barlow's Freedom of the Press Foundation, which by now
had added Snowden to its board, dropped Jake as an unpaid advisor.
Noisebridge, a warehouse-sized San Francisco hacker space Jake had
cofounded, said he could not come back.

Jake's early defenders included some Tor node operators, EFF
cofounder John Gilmore, and Daniel J. Bernstein, an antigovernment
cryptographer who had helped loosen export rules with Cohn's legal
help years before. Most cautioned against rushing to judgment without
legal process. Now a professor in the Netherlands and a major figure
in spreading non-NSA-backed encryption, Bernstein kept Jake on as a
graduate student.

The revelations were especially painful for cDc, which had built
Jake's credibility with other hackers. His conduct underscored the male
domination in security generally and in the hacker social scene in particular. Worst, Jake embodied the dark side of cDc's formula, wielding
a media-savvy, boundary-flaunting personality that could drive awareness while also feeding a rapacious ego.

What had made cDc special was shared values despite different viewpoints and areas of expertise, and that had been shattered.
"Those of us who knew Julian back in the day always knew he was
kind of a shit. I personally was always dubious of WikiLeaks largely
because of that," Paul Leonard said. "The reasoning all goes back to
the core of cDc, and furthermore was why Jake Appelbaum hit us in an

unreasonably painful way, which is that to an extent cDc functioned as something akin to a family unit."

[x x]

cDc could have said nothing. It was not as famous as it had been a decade earlier, and many of the articles about Jake wouldn't mention his affiliation with it. To the group's credit, its far-flung members scrambled even before the anonymous website appeared or Tor elaborated on its one-sentence announcement of Jake's resignation. Jake was still on the cDc mailing list, so the discussions had to happen elsewhere, including in smaller email chains. Luke alerted Kevin and Laird to early references on Twitter about rape accusations. Christien Rioux also wrote to Misha Kubecka. The concern was followed by caution.

"It's bad news, but I would definitely want to see more evidence than just some random dudes on twitter before we took any action internally," Luke wrote. Misha spoke from the gut: "Fuck. What is up with WikiLeaks people and rape?" After Christien spotted the anonymous accusers' website and passed along the link, Laird weighed in Saturday morning, saying that he had known that women had been trying to gather evidence of rape against Jake and that he had heard some "unsavory" stories of sexual conquests. "He can be a complete dick," Laird wrote. "I have my own experiences with this when I hosted him in India and he pulled some dumb stunts." Luke added Adam O'Donnell to the thread and suggested they seek out Jake's side of the mess.

The news stories started showing up on Sunday, and a friend of the group, Nick Farr, wrote publicly about being threatened by Jake and his supporters. That happened after Farr obliged Jake's demand that he cancel a five-minute talk during the Chaos Computer Congress's open-mic session by someone claiming that Jake was a US intelligence plant. Farr refused to hand his correspondence with the would-be speaker over to Jake. "Every night, I came back to my hotel room, a typewritten note on my pillow stating, 'Don't make us use extreme measures. Hand it all over.'" Farr said he contacted people he thought he could trust to explain what he was doing, but they all told him to find a compromise. "You can't dialogue with a sociopath," Farr wrote. "What's worse is when people you consider your trusted friends take the sociopath's side."

That was enough to push Laird toward making a public statement, and Adam seconded. Without having heard back from Kevin, Luke asked Misha to remove Jake from the mailing list so that they could out a proposed decision and statement to the full group. Finally, late Sunday night, Kevin showed up and said he wanted to quietly remove every trace of Jake from cDc sites, including the alumni roster. "I'm very sorry for my part in accepting this guy. That was dumb," Kevin wrote. "What I've come to realize is that personality matters 100% more than skills for this stuff. Whether any of these allegations are true or not, he doesn't fit in with us." The group convinced Kevin that they needed a public distancing. They collaborated on what would be its most serious public statement in more than a decade, then posted it to the cDc home page and to the then-open cDc Facebook group, where many of the members and fans shared information.

"Like much of the hacker community, we were troubled to hear the allegations of sexual abuse, manipulation, and bullying leveled against one of our members, Jacob Appelbaum, A.K.A. ioerror," it began.

> We're also aware that the Tor Project is conducting an internal in-vestigation, and encourage anyone with relevant testimony to come forward. For some, it won't be easy. There can be shaming or humil-iation, or the fear of not being believed. It is also our responsibility to create an environment where people feel safe to come forward. We have always stood for freedom of speech and expression, which sometimes necessitates the right to anonymity. This is something that victims of abuse often require. We stand by their right to be anony-mous. Others, like our friend Nick Farr, who decided to go public with his own difficulties, deserve our respect and support. Everyone will do this in their own way. We know that it may be scary, but we also encourage victims to contact their appropriate local authorities.

We understand the complicated relationship we all have with law enforcement, but there is a time and place for government intervention. If the most extreme of these allegations are true, they should be addressed in a court of law, and dealt with appropriately.

CULT OF THE DEAD COW is known for a lot of things, but treating people horribly is not one of them. If communities are to thrive and remain relevant we have to do some housecleaning from time to time. As we have become aware of the anonymous accusations of sexual assault, as well as the stories told by individuals we know and trust, we've decided to remove Jake from the herd effective immediately.

In a personal post on Medium, Laird said he hoped the ouster would help educate others about systemic sexism in hackerdom, exacerbated by a tendency toward rule-breaking, distrust of legal authorities in reporting transgressions, and some excessive scenesterism: "There's been a lot of looking the other way in the hacker community when powerful people overstep the bounds, and that has to stop."

It didn't take long for that wish to start coming true. As the broader antiharassment movement known as #MeToo built up steam in the fall of 2017, the hacker community rose up against other accused predators. Even Cap'n Crunch, John Draper, who had haunted hacker cons from the days of HoHoCon, was finally outed for pursuing underage boys and banned from gatherings. A Draper spokeswoman denied his seeking sex.

At least Jake was gone from cDc before the election of 2016, when his association with WikiLeaks would have been indefensible to everyone in cDc. WikiLeaks would be a central, partisan player in helping elect Trump, who lavishly praised it on the campaign trail. Emails stolen from the Democratic National Committee by Russian operatives were gleefully published by WikiLeaks as the Democratic convention was getting under way, when they could be dumped with maximum impact. Hours after Trump's campaign was blown off course by the publication of a video in which he bragged of grabbing women "by the pussy," WikiLeaks muddied the day by beginning to roll out stolen emails from Clinton campaign chairman Podesta. Long-promised leaks about Russia, meanwhile, never materialized. And Assange repeatedly tried to throw off suspicions with misinformation, denying that Russia was a source and hinting that a DNC staffer was one. In the summer of 2018, the special counsel's indictment of twelve Russian military intelligence officers would quote the emails between WikiLeaks and its real source, a Russian-created persona calling itself Guccifer 2.

Jake and Assange were far from alone in draping themselves in morality while serving other causes. Instead, they were just the most prominent exemplars. From 2016 on, a substantial amount of purported hacktivism would be something else in disguise.

CHAPTER 11

MIXTER, MUENCH, AND PHINEAS

WHILE JAKE APPELBAUM provided one example of the Cult of the Dead Cow's negative influence, he was not the only one. Edward Snowden had pulled back the veil and showed the symbiosis between Western intelligence agencies and big technology companies. cDc blood had infused both sides of that relationship, and both had lost moral luster. But soon cDc's descendants would be playing on all sides of an increasing complex struggle among spies in many countries, their technology suppliers, and the enemies of those suppliers—both those opposed for moral reasons, like the Citizen Lab, and those opposed for geopolitical reasons. Then, too, there would be anonymous vigilantes with motives hard to discern. For the most part they remained hidden, protected either by the technological sophistication of the very best hackers or by the tools provided by a nation-state.

The root cause of all this mess was the deeper integration of the nearly indefensible internet into all major economies during the tech industry's amoral drift of the 2000s. As that happened, it was inevitable that big governments would use security weaknesses to their advantage. It did not have to follow that they would ignore basic defense research, but they did that too. So what cDc had called out as a looming disaster at the turn of the millennium—shoddy software, uneducated buyers, and disengaged officials—had gotten much worse over the next decade. Instead of acting, perhaps in concert, to improve the security of what was driving economic growth for everyone, governments were supporting a dark market for knowledge about specific software flaws and techniques for exploiting them in order to spy. For some governments, the top targets were human rights advocates, journalists, and minority-party politicians. The people Laird Brown and his ilk had set out to protect were now in a much worse position than a decade before. "When I was young, there was something fun about the insecurity of the internet," complained Signal inventor Moxie Marlinspike. It opened up possibilities for anyone inventive enough to take advantage, despite their underdog status. Now "internet insecurity is used by people I don't like against people I do: the government against the people."

Of course, many of those who ended up supplying tools to the wrong people began with good intentions, including an early supporter of Laird's Hacktivismo project named Martin Muench. As German hacker Kemal Akman, called Mixter, was writing the Six/Four proxy system for Laird, the latter-day cDc member had landed a job at a Munich start-up called Ciphire Labs, which was trying to develop an encrypted email system. Kemal helped get Laird a job there as well, and he looked for colleagues at the company who might volunteer for Hacktivismo. Kemal thought he'd found a perfect candidate in Muench, an intense and brilliant teenager. Kemal added him to the Hacktivismo mailing list, which had about twenty steady contributors and ten times as many quiet readers who lurked and learned. Kemal also introduced the young man to others he knew in the Berlin scene centering around the Chaos Computer Club.

"Martin was an idealist," Kemal said. "He had my full trust." But Muench "wanted to be a rock star," and that also influenced his trajectory. Muench told Kemal that he wanted to help the police catch the worst of the worst, the makers of child pornography, and he left Ciphire to go work on software that he said would help. Because he had recruited and mentored Muench and introduced him to human rights–oriented hackers around Berlin, Kemal felt guilty for what happened next. "I put Martin on there. I am partly responsible for his career," Kemal said. "I personally found him a bit strange." Laird said he still regards Muench as a friend, one who lost control of a project due to a struggle inside the company he started, not because he had chosen an immoral path.

Whatever the reason, Muench's system became the next flashpoint in the fight over hacking, security, and privacy. As with WikiLeaks, the debate would go beyond professionals and engage the media and general public in a discussion about the balance of power between governments and average citizens. Though hunting from behind blinds, Russia has been driving a significant chunk of that debate, probably with Muench, and definitely with the dumping of hacking tools used by the National Security Agency. While cDc had started blending political motives and security work with the Hong Kong Blondes, two decades of increasing geopolitical influence on hacktivist causes since then have made it hard to tease apart the real actors and purpose behind many public hacks.

Muench's modest program developed into spyware called FinFisher or FinSpy. In Kemal's thinking, Back Orifice 2000 inspired the project, and Muench might have adapted some code from that open-source effort. But Christien Rioux was aiming only at Windows computers. Muench's FinFisher targeted Windows and Apple computers, Android and Apple phones, other devices, and most operating systems. There were other differences too. Anybody could use Back Orifice 2000, but those users needed to find a working exploit, or a gullible victim, to get it installed. The company selling Muench's FinFisher, called Gamma Group, also provided the tricks to get it installed on the devices. Muench headed product development and Gamma's Munich office. Gamma also had a headquarters in the UK and affiliated companies in Singapore and other countries, which ostensibly sold only to established government agencies.

Kemal heard about Muench's connection to Gamma in 2008. In 2011, activists infiltrated the trade shows known as the Wiretapper's Ball and emerged with a sixty-page Gamma catalog. "FinFisher is the cutting edge offensive IT intrusion portfolio on the market today," it declared. The portfolio included impressive spy programs aimed at smart phones. Those were very hard to detect, could be operated re-motely, and would not only capture voice calls and electronic contact lists but also turn the phones into constant surveillance recorders. The same year, during the Arab Spring, Egyptian rebels discovered a similar pitch was given to the brutal State Security Investigations Service. Gamma said that the deal hadn't gone through, that it com-plied with export laws, and that it only sold to governments targeting criminals.

But activists suspected repressive regimes, including those un-der widespread sanctions, like Sudan, were using FinFisher against law-abiding dissidents. In 2012, Bloomberg News obtained suspected infected emails sent to Bahraini activists and handed them off to the Laird-inspired Citizen Lab. A Citizen Lab team led by a Google se-curity expert dove deep. For the first time, they established that the infections were connected to Gamma, how the infections worked, and that data from the victims would be sent to the Bahraini government's telecommunications company. Citizen Lab found FinFisher servers in dozens of countries, including the UAE, Ethiopia, and Vietnam, where bloggers were being targeted. Technological tricks used by the com-pany included intervening in software update processes and using ex-ploits for Adobe's Flash video software.

Two years later, someone hacked Gamma, badly. The hacker opened a parody Twitter account, @GammaGroupPR, and tweeted links to stolen files with source code, client lists, and other damag-ing information, including a chart showing that the largest number of visitors to the company's customer-support pages had shifted from the Netherlands, France, and China in 2009 to China and the US in 2014. The tech press had a field day, activist researchers rejoiced, and non-profits filed complaints with authorities that severely hurt the company.

In 2015, @GammaGroupPR came back to life to announce that it had also hacked Gamma's best-known rival, Italian company Hacking Team. As LulzSec had with HBGary Federal's Aaron Barr, the hackers delighted in pointing out the company's poor security. Once more they dumped source code, client lists that showed apparent sanctions violations, and embarrassing emails. Hacking Team tools had been used against Ethiopian journalists and other innocents, including some inside the US. The person controlling the @GammaGroupPR account, who referred to himself as Phineas Fisher, said in a later interview with VICE that he had gone after both companies out of moral outrage. "I just read the Citizen Lab reports on FinFisher and Hacking Team and thought, 'That's fucked up,' and I hacked them," he explained. "Hopefully it can at least set them back a bit and give some breathing room to the people being targeted with their software." In that interview, conducted over electronic chat in July 2016, Phineas used informal English and alluded to the Antisec movement from years earlier. Describing himself as an "anarchist revolutionary," Phineas published tutorials and a manifesto encouraging others to hack their oppressors.

In another interview a month earlier, Phineas admitted to hacking a Catalan police union, in the Spanish region that includes Barcelona, and posting the home addresses of more than five thousand officers. He called it a "small strike against power" and denied being Spanish or speaking either Spanish or Catalan. All the same, that very local target fueled speculation that Phineas was a politically minded hacker from the region.

Phineas's stunts took the original Antisec movement and HBGary breach in exactly the direction that previous hacktivists who were willing to break the law would have gone. He used his knowledge of how the world really works to make it harder for technology to be applied for oppression. After Phineas came leaks of purloined material from Cellebrite, an Israeli tech forensics company that breaks into phones for law enforcement, and from the makers of FlexiSpy, spyware used by parents to track children and by romantic partners to snoop on each other. (Cellebrite had been reported by some publications as the company that helped the FBI finally break into the Apple iPhone of a terrorist who killed public employees in San Bernardino, California, after Apple refused to do it.) The FlexiSpy hackers paid tribute to Phineas and published an updated security and how-to-hack guide for fans. "If you're a hacker, hack back," they wrote. "If you're an ordinary person, stay safe. Watch how things progress, and see what people are saying about how to detect FlexiSpy and protect yourselves. . . . If you're a spouseware vendor, we're coming for you. Stop, rethink your life, kill your company, and be a better person. Otherwise, you'll be seeing us soon." Gabriella Coleman, the Anonymous chronicler teaching at McGill University, called the trend the birth of "public interest hacking," and it is likely that at least some of what grew to more than a half-dozen spyware breaches stemmed from moral objections to the vendors' conduct.

[x x]

All the same, it is worth taking another look at instigator Phineas in the wake of the hack of the Democratic National Committee and publication of NSA tools. The basics of the DNC breach and others against Democratic Party officials during the 2016 US elections have been clearly established by US investigators, including those working for special counsel Robert Mueller. One DNC breach came shortly after the publication of a Pulitzer-winning series of articles about the leaked Panama Papers, which showed that Putin's friends were stashing billions of dollars overseas. Putin blamed Clinton for the underlying leak of an offshore law firm's files. Given that US intelligence had indeed debated exposing Putin's corruption, he may have been right that it was a CIA operation. Though Assange tried to cast doubts about who provided WikiLeaks with stolen emails, Russian intelligence clearly drove the attack on the DNC and related hacks. The persona Guccifer 2, which shopped some stolen DNC data while claiming to be Romanian, once forgot to use a virtual private network to connect and revealed his true location at the GRU, Russia's military intelligence service. Russia also arranged for the publication of emails and other documents by WikiLeaks and others.

The major NSA breach has not been tied up as neatly. In August 2016, just weeks after Phineas stopped bragging, a group calling itself the Shadow Brokers appeared on Twitter and began dropping not only vulnerabilities in Microsoft's Windows, Cisco routers, and other programs but also working exploits, all of which had been held by the NSA. Most of the information came from late 2013, after Edward Snowden had left the agency, meaning that there was another mole, or a hack of agency hardware, or a careless employee who had been hacked. Shadow Brokers kept going for months. Some of the tricks it disclosed were then used by others, including the presumed North Korean distributors of badly crafted ransomware called WannaCry, which shuttered hospitals and other facilities around the planet in 2017. Eventually, two NSA employees were charged with bringing classified files home. At least one of them had been running Kaspersky antivirus on his personal computer.

That was cause for special concern, because the Israelis had broken into Kaspersky's networks in 2015. Inside, they had seen that the software was used to search for classified US documents, and they had warned the Americans. The consensus in the intelligence agencies was that the Russians had obtained at least some of the Shadow Brokers information in that manner. The disclosures badly hurt Kaspersky, which had enjoyed a remarkable run publicly exposing high-end US malware, starting with the Stuxnet virus, which had knocked out Iranian nuclear centrifuges. Kaspersky admitted it had taken some secret files from a US government employee, though it claimed that it had deleted them. The US banned it from federal government use.

The Russians had the motive to steal US hacking tools, the means to do it, and the opportunity. Russia was also one of the few suspects with so many of its own tools that it could afford to dump those of the US instead of hoarding them for its own use. The timing is particularly interesting, since the NSA dumps began in August 2016, two months after the DNC breach was disclosed. Russia created chaos and distraction inside the agencies best able to find the source of the DNC hack and strike back, helping to paralyze the Obama administration and mute its response.

With that history in mind, it is worth revisiting the identity of Phineas Fisher. Missing from virtually all of the mainstream media coverage was the fact that Gamma Group and Hacking Team generally did not sell to Russia or its closest allies. They sold hacking tools to the West, and Phineas stole them and dumped them in public, just as the Shadow Brokers would do starting weeks later with the NSA. In addition, Gamma Group held special interest for Kaspersky. Two former Kaspersky employees told me that the company had lifted inactive code from a Gamma computer after someone there foolishly installed their antivirus software.

And then there are the matters of Phineas's choice of other targets and what we now know to be Russia's strategy of sowing division in the European Union, in the US, and in other strategic countries. An attack on the Catalan police union would fit with pitting regions against the central Spanish government, which emerged as a Russian goal in 2017 when the Catalan government defied a Madrid court order and held a referendum on seceding. After Spain ordered the Catalan leader removed, the loyalty of the police was of enormous importance.

It would be surprising for a highly skilled, willfully lawbreaking, and morally driven hacker to take down both Gamma Group and Hacking Team and still get seriously involved with Spanish political issues. At a minimum, you would expect someone with that combination to be Spanish. But that is not all Phineas did. He also hacked the data of innocent Turkish citizens during a period of confrontation between Russia and Turkey and made that data public. Though this context was missing from most of the hack's coverage, Russia and Turkey had been in an escalating confrontation since Turkey had shot down a Russian plane, killing its pilot, in late 2015. Over the next half-year, Putin increased pressure on Turkish president Recep Erdogan with sanctions on Turkish food imports and a ban on sales of Turkish tourist packages to Russians. At the same time, Erdogan was losing popularity in the West as he cracked down on the media and activists. Meanwhile,

Russia and Turkey were pursuing different goals in neighboring Syria, Russia's client state. Erdogan had to choose between Washington and Moscow, and he eventually opted for the latter. Even though the downed plane was believed to have been in Turkish airspace, Erdogan folded and wrote to Putin in June 2016: "I once again express my sympathy and profound condolences to the family of the Russian pilot who was killed, and I apologize to them."

Erdogan had planned to purge the military, and that triggered a coup attempt in July 2016, it emerged later. Russia was the first outside nation to condemn the coup, suggesting that the alliance had been firmed up beforehand. But many pieces were in play at once, and it would make sense for Russia to have been interested in weakening Erdogan's hand by exposing his party's followers to scrutiny, in the same way that exposing the Catalan police officers' personal information could have been useful in stirring the pot. Perhaps Russia was betting on both sides of the Erdogan-military conflict, so that whoever emerged victorious would be in its debt. In any case, it would make more sense for Russia to publish such information than it would for a politically minded hacker in Spain or elsewhere.

Phineas's explanation for what he was trying to do and what went wrong, on the other hand, makes little sense. "I hacked AKP (the ruling party in Turkey) because I support the society [Kurdish] people are trying to build in Rojava and Bakur, and they're being attacked by Turkey," Phineas posted in July. He then added a complex story for why sensitive information was published about ordinary people. According to Phineas, he had hacked into the party's servers and shared a historical file of emails with people in the restive regions, asking them what he should do with the access. The emails themselves were not interesting. They included people asking for potholes to be fixed or for help finding work. There was nothing from Erdogan or his inner circle. Then, "there was a miscommunication between some of them," Phineas wrote, and one of the people gave the dump to WikiLeaks. He said that even though the person who had relayed the files realized the mistake and asked WikiLeaks not to publish, it did so anyway.

But Phineas then published more files himself, including a database of ordinary AKP members and, worse, a database of almost all the adult women in Turkey, along with cell phone numbers and addresses for many of them. Those databases were copied and reposted by people like UK security activist Thomas White, who tweeted as @CthulhuSec and had won a measure of controversial fame by posting the fruits of many large hacks. WikiLeaks tweeted links to those databases, which allowed millions of women to be reached by stalkers, further angering previous admirers of Phineas, such as Electronic Frontier Foundation activist Eva Galperin. "Who's that behind the not-so-great leak of Turkish emails?" Galperin wrote on Twitter. "It's @GammaGroupPR, whose previous work I have enjoyed." Three months later, White stopped posting links to leaked data, complaining that the motives of the hackers had gotten more crass. Three months after that, Phineas told VICE he was retiring his moniker and that he would take a break from all hacking.

So now we have a hacker who is extraordinarily skilled, ethically driven, and broad enough in his thinking to go after both the rank-and-file regional police in Barcelona and the Turkish ruling party, yet sloppy enough to expose the phone numbers of millions of women in a patriarchal society to the general public, along with those of ordinary party members just as they become uniquely at risk if exposed. It seems unlikely. Even without the relationship with WikiLeaks, an equally logical explanation would be that Phineas is a Russian intelligence project. Indeed, that was Washington's private conclusion. Within US intelligence, "it's generally assumed to be Russians," said Jim Lewis, a well-connected longtime senior State Department official and negotiator on global internet issues. "It's consistent with Russian activities in other areas."

If the Russians did try to ruin Gamma Group and Hacking Team, they had their own tools for spying on citizens and enemies and were merely making life harder for governments in the West. That doesn't necessarily mean that those companies didn't deserve exposure. Kemal, for one, didn't hesitate before applauding the leak, even if it came from the Kremlin and hurt his old friend Muench. "I'm really happy about it," he said of the exposed Gamma tools. "They should be leaked, and they should be burned."

[x x]

Even if Phineas isn't Russian, a look at the bigger picture is warranted. We have to accept that hacktivism is often polluted by geopolitics—as in fact, it was with Laird—and that such influence can be impossible to detect. If that weren't alarming enough, there is a deeper realization. The great powers of the world contest each other in public and in secret, using arms and money, diplomacy and spying, false activism and public relations. At the same time, most governments have similar interests against their own people. None of them want their citizens to be able to communicate in secret, not even the United States. In 2018 the FBI was still railing against the ability of people to use encryption that vendors cannot break, and congressional allies were still threatening legislation to outlaw such security.

Kemal saw the trend toward greater government power against the individual as so dispiriting that in 2011 he left the security industry for years. Like others in cDc, he felt the best remaining hope for preserving individual freedoms lay with the biggest vendors, like Apple and Google, who could in theory play the major governments off one another and protect users in the process, and with private start-ups like Signal that feel they are chasing things more important than money.

Apple was clearly a battlefront. It was home to @stake veterans Window Snyder, David Litchfield, and Rob Beck, along with many more cDc kindred spirits. Some of them helped stave off the FBI's attempt to force it to crack the San Bernardino iPhone. Apple argued that the government could probably find a way to break the phone on its own and that making it write a new program would be compelled speech, which has been found unconstitutional. The FBI was losing when it suddenly found an unnamed contractor with a zero-day that could do the job and dropped the case. Google was another war zone stocked with cDc members and admirers. It had realized the NSA was the enemy after Snowden documents showed the agency had been breaking into its networks overseas, where it did not need court approval. Google moved to encrypt far more deeply, even if it maintained the ability to recover all users' emails. The two companies also fought against proposed government-mandated back doors and bans on end-to-end encryption, which by 2018 were popping up around the globe.

There was still fighting to be done inside the big companies. But leading lights in the encryption fight were also spending more time helping the start-ups. Others were beginning to think more about the meaning of free speech when the immediate problem in many countries was not the inability to speak but the propensity to get drowned out by manufactured voices directed by governments and big economic forces. Laird and the others in cDc were appalled at the likes of Gamma Group and sorry to have played any role in Muench's rise. But while they may have cheered Phineas on, they were not interested in breaking the laws themselves. As the hacktivist battlegrounds evolved toward hacking, leaking, and information warfare, they had to find other ways to help.

After Laird returned from India to Germany, he went back to work for the former CEO of Ciphire, the now-failed encrypted email provider that had also employed Kemal and Muench. The ex-CEO, Errikos Pitsos, had an idea for a platform for serious debate he called Kialo. The software guided discussion by showing decision trees that listed which followers had agreed with which points. Moderators rejected unhelpful comments. Pitsos funded the project himself, aiming to create a "collaborative reasoning tool," and Harvard and other universities tried out private versions for classrooms. It wasn't going to get rid of the bots and trolls on Twitter, but at least it was positive. On the side, Laird began writing a book on information warfare.

Some sympathetic to US cyberoperations, including Mudge, also saw a clear ethical case for authorized offensive work. They decided that hacking in order to spy, to prepare the battlefield in case of further conflict, and to conduct highly targeted destructive attacks, as with

Stuxnet, was vastly preferable to sending in bombs and troops. Others in cDc, looking at the mixed motives as geopolitical priorities ascended, opted to go back to basics on defense. By making the internet safer for everyone, they could chip away at the unfair advantage the net had been giving to attackers since the beginning.

> CHAPTER 12

> MUDGE AND DILDOG

PEITER ZATKO, KNOWN to even close friends as Mudge, was not the most engaged executive at @stake, even though he was the lead creator of the pioneer hacker consulting group. The most famous member of the Cult of the Dead Cow was elsewhere much of the time, fighting his own demons and, after 9/11, America's demons as well. What he saw made him very afraid. Mudge knew as much as anyone about the basic failings of tech security and about their root causes. The internet's inventors built it on trust and it got loose in its test version, before Vint Cerf and his team could come up with reliable security. It still ran that way.

All software has bugs, some of which can be exploited. Layering software on software makes it less secure. The software vendors had all escaped legal liability for poor craftsmanship and had little incentive to devote significant resources to making their products safer. (This hard line on liability has only begun to fray in 2018 in extreme cases, as with deaths blamed on automated vehicles' programming.) Regulation ranged from nonexistent in most commercial markets to negligible in industries such as financial services, health care, and power distribution. All of which meant everything was unsafe and would only get less safe as the economy grew more dependent on technology.

This was classic market failure, compounded by political failure. One could debate the largest causes of the political failure, but they included the capture of the regulators by industries that did not want to be regulated, the dominant pursuit of short-term business gains by short-term business executives, and the failure to distinguish when private companies should be responsible for their own defense and when the federal government needed to step in. That last was nontrivial, since the same techniques could be employed by criminal hackers, fending off whom would generally be considered a corporate responsibility, and nation-state spies, who would generally be considered a Homeland Security or FBI responsibility, with backup from the Department of Defense. Even if those lines were clear, what do you do about criminals who work for spies, or spies who moonlight as criminals? Congress's inaction loomed large. But without blood on the streets, Mudge held little hope of that changing anytime soon.

In 2003, as largely Russian organized crime groups took the leading role in spreading computer viruses for spam and extortion, Mudge saw that the big picture was about to look a lot worse. He figured the best way to help was to go to the place that had the best understanding of the problem, the most power to deal with it, and the greatest responsibility: the federal intelligence agencies. Given his sketchy associations and general antiestablishment attitude, it would have been draining to apply directly at the CIA or NSA. But Mudge could at least start where he was a known quantity, and where he had geographical and employment buffers from the people wearing braids and stars on their uniforms.

A year after Mudge's top government sponsor, Richard Clarke, resigned from the Bush White House, Mudge rejoined BBN Technologies. Starting in 2004, he worked at BBN on research and development for US intelligence agencies, and he trained people who would become the core of the NSA's elite hacking unit, Tailored Access Operations. Over the next six years, he worked on a lot of things he can't talk about. "I think domestic lives have been saved as my ideas went operational," Mudge said. He told me that lives in the Middle East were also saved because his tools were used instead of bombs.

In 2010, the new head of the Defense Advanced Research Projects Agency asked Mudge to come in-house and lead the agency's cybersecurity efforts. Mudge had thought about DARPA before, but he hadn't been enthusiastic about the agency's prior leadership. The new boss, Regina Dugan, he liked. And DARPA, founded in 1958 in response to Russia's stunning Sputnik satellite, had the coolest mission in government: "the creation and suppression of strategic surprise." Like many positions inside DARPA, the post was for a fixed three-year term, during which he would award grants for offensive and defensive breakthroughs in security. But the opportunity was incredible. This agency had steered the creation of the ARPANET, which became the modern internet. "I obviously wanted to make sure the things I depend on, that my family and friends depend on, are secure," Mudge said. "I also owe a lot to my country. A lot of countries would not have allowed me to influence the intelligence community and the Department of Defense, hopefully in ways that have them make less stupid mistakes."

Mudge's personal slogan had long been "Make a dent in the universe." Now he called in a dozen of the smartest hackers he knew to help figure out how. He told them to be ready to discuss where the security industry was failing, what they as researchers were angriest about, and what DARPA could do to help. They convened in a bland Arlington, Virginia, building that housed the massive intelligence contractor Booz Allen Hamilton, the company that would employ Edward Snowden. Mudge's call brought out "a bunch of misfits," said Dug Song, who was among them. The group included @stake veterans Dave Aitel, now running zero-day seller Immunity Inc., and Dino Dai Zovi, a former federal labs researcher and chief scientist at government zero-day supplier Endgame. Also there was sometime intelligence contractor H. D. Moore, who had created Metasploit, a penetration-testing tool that used vulnerabilities as soon as they were disclosed, often within a day. Ninja Strike Force stalwart and intelligence contractor Val Smith came too.

Mudge convened the meeting by telling them that his DARPA slot had given the entire hacking community, at long last, "a seat at the table." Now, he said, "let's not waste this opportunity." As they brainstormed priorities, Song asked about something different: a change in process. DARPA funded the big guys—defense contractors, other major corporations, and some university departments. Those operations knew how to navigate the paperwork, come up with slick pitches, and leverage their previous work. This left out talented small teams and individuals who had great insights from being hands-on hackers and no idea where to go from there. The son of a liquor-store owner, Song had used a small-business grant to start Arbor Networks. He said DARPA should go small as well, and Smith agreed.

Mudge had spent enough time around government to realize they were right, and he convinced Dugan. "The process itself was an impediment," Dugan said. Mudge announced the Cyber Fast Track not long after, the first program at DARPA aimed at giving small amounts to small teams, instead of large amounts to large ones. Mudge funded nearly two hundred proposals, all of which let the researchers keep their own intellectual property. Among the recipients was Moxie Marlinspike, whose invention Signal would come years later, and Charlie Miller, who studied flaws in near-field communication as those protocols were getting embedded in more smartphones.

At Def Con in 2011, Miller was presenting a near-field talk and bumped into Mudge, who was also speaking. Miller told Mudge some of the things he was interested in and asked if DARPA would buy him a car he could hack. "Submit and find out," Mudge said, so Miller did. He got the car and hacked away. Building on that work later, Miller hacked a moving jeep being driven by a *Wired* reporter, prompting a mass recall and drawing global attention to the safety issues of computerized vehicles. The initial equipment and the money was one thing. But DARPA's backing became even more important when a car company, upset at Miller's revelations, threatened to sue. Mudge warned them that if they did, the Pentagon would join the suit on Miller's side, with a significant number of well-trained lawyers.

"Those grants also provided a certain amount of legitimacy to the research that really helped when people were having objections," Miller said. "There are lots of research projects you see around now that would have never existed without those CFT grants, including the car hacking we did." Everyone at the Pentagon wanted to get the papers explaining the research. But before they could get the briefing books, they had to sit through a demonstration by the hackers themselves, so they really understood them. In the years that followed, other areas at the Pentagon began mimicking the fast track Mudge developed.

[x x]

Mudge did much more than streamline the way the federal government acquired good ideas. He also tackled a fundamental problem with the way the government, and everyone else, evaluated security. For decades, no one had come up with a reasonable way to estimate the worth of security products, which draw attention mainly when they fail. Likewise, DARPA couldn't figure out a logical basis for determining what to fund. "We are not going to approve a single new project until we do the deep strategic work," Dugan said. She insisted that Mudge and his boss, long-serving DARPA software chief Dan Kaufman, find a new way of looking at the issue.

Mudge and Kaufman came up with what they called the Cyber Analytic Framework. The major concept: as predictable complexity increases, the defenders' job gets harder more rapidly than the attacker's job does. To illustrate the problem, Mudge used the common language of Washington, a slide deck. The most eye-popping chart showed that the average advanced defense software had bloated to contain 10 million lines of code over the past decade. The average number of lines in malicious software, meanwhile, had held steady at 125.

Since every thousand lines of code has one to five bugs in it, that meant big security products were making the situation worse. DARPA needed to seek simple and elegant approaches instead. "It was a clear articulation of trend lines," Dugan said. Mudge began asking defensive grant applicants whether their approaches were tactical or strategic, how their project would increase or decrease the overall attack surface being defended, and how they would beat it themselves.

The Framework approach became the basis for DoD spending beyond DARPA, and it got DARPA some money that otherwise would have gone to Cyber Command, one of several things Mudge worked on that annoyed Cyber Command head and NSA director Keith Alexander. Mudge didn't mind that at all. Alexander had presided over a massive expansion of global and US surveillance, as well as a culture that produced several whistle-blowers and leakers while allowing employees to be hacked.

Mudge loved betting on promising ideas, but he also considered it his duty to strangle bad ones in the crib. While still an outside contractor, he decried a product that automated some "active defense," the industry term for measures that range from blocking suspicious connections to disabling the computers used by an attacker. Though hacking back tempts targets that feel powerless relying on the government, most intelligence professionals think it is a bad idea that would lead to chaos and perhaps an unintended war. Automating that "is a terrible idea, because then an outsider can make you do things," Mudge said.

Mudge also expended considerable energy arguing against demands for back doors in encryption. Intelligence and military officials said that back doors worked well in their offices—that access was logged and controlled and that abuse was rare. But those were closed systems, where the people in charge could completely govern the environment. Out in the regular world, configurations get looser and privileged access leaks.

Mudge didn't stop telling the truth just because he was at the seat of great power. It probably helped that his position would end after just three years, so officials expected less sucking up. Mudge briefed the Joint Chiefs and the secretary of defense, helping them understand when one of the armed forces or a contractor was claiming an improbable capability in a turf or budget fight. "The Joint Chiefs and the Pentagon would call me in because I didn't have a horse running, and I was able to explain to them ground truth," he said.

Mudge remained iconoclastic. Amid widespread outcry over the constant breaches of American defense vendors by other nations, Mudge observed shortly after leaving DARPA that contractors had a perverse incentive to allow their weapons systems to be stolen. Once

that happened, Mudge mused at Black Hat, they could ask the Pentagon to pay for a new and improved version of their system that was not
yet in enemy hands. "Game theory is a bitch," he said.

Yet Mudge managed to play the inside game well. DARPA always
sent off its creations to new homes within the Pentagon or intelligence
establishment where they would best develop. With Alexander and others predisposed to dislike much of what Mudge had handled, he sometimes engaged in subterfuge, handing off to a midlevel operative who
could remove evidence of a project's heritage. At one briefing with the
deputy secretary of defense, Alexander explained that he had five "silver bullets" that he could deploy in cyberoperations. "Three of those
are mine," Mudge thought with satisfaction.

Mudge got the Pentagon to stop seeing hackers as the natural enemy. In fact, Mudge showed that people who grew up knowing exactly
where the line was were habitually more careful about not crossing it
than people constantly protected by their uniforms, bureaucracy, and
lawyers. During one discussion at a large agency that was witnessed
by Kaufman, an employee asked Mudge if the agency could just hack
into a system in order to get information Mudge was deducing. "Absolutely, you could do that," Mudge told him. "But just suggesting that is
illegal, and it's wrong." Even within DARPA, Mudge provided a moral
compass.

[x x]

In a fortuitous bit of timing, Mudge's scheduled exit from the government came in April 2013, two months before Snowden's disclosures
turned the NSA and US intelligence into global punching bags. On his
way out, Mudge accepted the secretary of defense's highest award for
civilian service. The citation said that Mudge's fast-track grants had
produced more than one hundred new capabilities, that his new method
for detecting cyberespionage had been placed into operation by intelligence agencies, and that he had improved the Defense Department's
ability to conduct online attacks.

Mudge followed Dugan to Google, where he worked on secret projects. The best known put a secure operating system on a memory card; the software would function properly even if the overall computer were compromised. Its features included an unchangeable logging system. The software would have been among the best possible defenses to the mass surveillance revealed by Snowden. Google did not release a finished version before Mudge left for a new venture: a nonprofit to examine code from binaries, the machine-readable instructions that programs give to computers, and score them based on standard safety features.

Mudge and his wife Sarah's Cyber Independent Testing Lab functioned like the labs at *Consumer Reports*, scanning for the digital equivalent of automatic brakes and seat belts, all without needing access to the source code. With money from DARPA, the Ford Foundation, and others, CITL showed that on a Mac's then-current operating system, hackers would have a harder time attacking Google's Chrome browser than Safari or Firefox. Mudge aimed to make a more detailed version of such scores into something like the mandatory nutritional labels on food, telling buyers enough for them to make informed choices that reflect their priorities.

Grappling with kidney cancer that brought back his post-traumatic stress disorder, Mudge saw the project through its first year, then handed day-to-day control to Sarah, a fellow veteran of federal contractor BBN. Mudge took a day job as head of security at internet payment processor Stripe, which helped pay the bills at the nonprofit. (A September 2018 investment round would value that company at $20 billion.) In his spare time, Mudge served as cybersecurity advisor to Senator Mark Warner, cochair of the Senate Cybersecurity Caucus. "Mudge has been extremely helpful in refining our understanding of software security, which informed our work on improving the security of internet-of-things devices, to take just one example," Warner said, referring to new classes of internet-connected gadgets such as security cameras and thermostats. Warner also served as the top Democrat on the Senate Intelligence Committee, making him the lead Democrat in

the congressional investigations of Russia's hacking to help Trump win
the 2016 election. It would be logical to think that Mudge's expertise
aided Warner there as well, though neither man would discuss it with
me. (Mudge had earlier advised the Democratic Party in 2016 to tighten
its security, he tweeted in 2018, but most of his advice was ignored.)

[x x]

The other great technical mind from cDc's golden era, Christien "Dildog" Rioux, wound up doing something technologically similar to the
work of Mudge's lab: deeply analyzing the safety of programs without
access to the source code. But he went a very different route, starting
with rejecting an opportunity to work for the government and ending up
doing something much bigger.

While with @stake, Christien spent a lot of time poring over binaries. Source code, which appears as it is written by the programmers, is a hundred times easier for the human eye to comprehend. But
it can also hide a host of ills. Looking at ones and zeroes, though, is
mind-numbing. So Christien wrote as many tools as he could to process
the binaries and tell him what they were saying to the computers. That
saved a lot of time while still allowing him to conduct what the industry
calls static analysis of the code. As Symantec sucked @stake deeper
into itself and made it harder to distinguish from the rest of the giant
company, Christien decided to create a start-up to fund his quest for
something of a holy grail—a program that would decompile all the binaries back into human-readable instructions for analysis.

From 2006, Christien served as chief scientist of the new company, called Veracode. He tapped Chris Wysopal, his colleague from
the L0pht and @stake, as cofounder and chief technology officer. The
business plan called for them to serve software customers instead of
the makers, like Microsoft and Oracle, where there were incentives to
scrimp on security. Once the master program worked well, Christien
reasoned, the buyers could convince their suppliers to let Veracode do
a safety analysis on the binaries. If they passed with flying colors, then

the suppliers would cite Veracode's approval as a badge of honor and recommend that prospective customers have Veracode do a new check on the most recent software version.

In theory, it was brilliant. In practice, it was a lot of work. "It was a five-year business plan that executed really well in ten years," Christien said. One early round of funding came from In-Q-Tel, the Silicon Valley venture firm set up to serve the needs of the US intelligence agencies and led by former @stake CEO Chris Darby. Darby believed Christien would make code much more secure, and he thought Christien should deploy it inside US weapons systems, making sure that the code controlling missiles and the like could withstand most attacks by hackers.

Darby arranged for Christien to visit an intelligence installation deep underground and give a demonstration of what Veracode could do. A senior officer of clandestine operations said hello, adding, "I'm a big fan of you guys from the L0pht." Christien thanked him. "What a nice guy," he thought. "He probably kills people." On a specially prepared laptop, Christien analyzed a blob of binary code that had been given him, perhaps a spying tool crafted by the agency. He let the program run during a lunch break and came back just as it spat out the results, describing what many pieces of the code did. Among other things, it detected a custom modification of a standard encryption algorithm. The polite killer was blown away. But the logistics of a major deal were daunting. Veracode could provide its program, but it could not be around to maintain it.

Darby wanted Christien to focus on optimizing the code for such deals anyway. But Christien figured that his main customers would end up being the federal government and a few close allies. "This would not be very lucrative for me, and it would have me working five hundred feet underground and never seeing the light of day," Christien thought. He didn't even want to go through the hassle of getting a security clearance. More importantly, "I want to have a bigger impact on the world, and I don't see it happening in the bowels of government."

Once Veracode decided to stay focused on the commercial world and Christien's team cobbled together a prototype of their master decompiler, he and Wysopal started calling old friends who were now inside the big software companies. That included Brad Arkin, an @stake veteran who by 2008 was a senior director for security at Adobe Systems, perhaps the vendor most criticized for omnipresent software flaws in all of Silicon Valley. "Everyone knows your Flash player is full of bugs," Christien told Arkin, promising to find all the problems. "We can do a scan in a month." Arkin agreed. But the code base was a mess on a scale Christien had never seen before. In addition to regular programming foibles, Adobe had incorporated obscure encoding systems so that it could display material recorded in all kinds of formats and show them on many different devices. It kept choking the decompiler. After a month went by, Christien declared that he would not shave until he was able to complete the Flash scan. That kept the pressure on him. But it still took an entire, brutal year, and his face itched like hell. "I hate Adobe," Christien said.

Pulling through it made Veracode's product much better. The company added big software customers, and by working through military contractors like Boeing, it could also serve the NSA and CIA. Veracode convinced software buyers to demand that their vendors allow Veracode to audit the binaries, which were stored on extremely secure computers. The first time through the wringer, most of the software providers hated it. But instead of blowing the whistle on those suppliers immediately for major weaknesses, Veracode would give them a couple of chances to improve, along with pointers about where and how to do that. Like many software and service companies, Veracode's sales went up and down, with extra volatility around the end of the quarters because of the commission incentives. After the company straightened that out, and with sales approaching $120 million a year, Veracode weighed going public. The alternative was selling itself to a company with deeper pockets that could bring Veracode to more customers. The latter ended up being a better deal, and Veracode sold itself in 2017 to CA Technologies, formerly known as Computer Associates, for $614 million. It was sold and resold in the following year, the last time for $950 million. Once installed in his new corporate home, Christien could spend more time on a side project called Hail-stone, which allows developers to test their code for security flaws as they write. While Veracode typically cost $10,000 a year, they could try Hailstone free. He quit Veracode entirely in March 2019.

[x x]

The largest proportion of Cult of the Dead Cow members wound up working at tech companies with people who didn't know their history. That included Luke Benfey, Paul Leonard, Matt Kelly, Misha Kubecka, and Kemal Akman. The previously outed Josh Buchbinder still works in security in San Francisco. John Lester is in Montreal: he worked for the maker of Second Life for years, then focused on electronic tools for interactive medicine and education. Dan MacMillan turned toward business, becoming a sales and consulting executive at big software companies. Glenn Kurtzrock, who had always wanted to put bad guys in jail, served as an assistant district attorney on Long Island for seven-teen years before starting a private practice in 2017. Carrie Campbell is a freelance researcher near Seattle. Cofounder Bill Brown teaches doc-umentary film. Cofounder Brandon Brewer, once known as Sid Vicious, is as straight as it gets: senior vice president of real estate–services firm Republic Title, based in Fort Worth.

Sam Anthony went to work as a programmer in a Harvard Uni-versity lab, then started graduate school there, working on biological models for computation. He earned a PhD in 2018. Along the way, he cofounded a self-driving car technology company, Perceptive Automata. Autonomous vehicles "are super good at knowing where the road is, how fast the car is going, whether something's a tree or a person," Sam explained. "They're miserably bad at solving the psychology problem of guessing what's in a human's head. The techniques we developed while I was doing my PhD are perfect for situations where you want ma-chine learning to do something where humans are incredible." Sam's company took video clips of pedestrians, showed them to humans, and asked such questions as whether the subjects were acting like they wanted to cross the road. He was using machine-learning techniques to teach computers how to understand people. By the Consumer Electronics Show in January 2019, Perceptive Automata could boast of investments by Toyota, Honda, and Hyundai.

Kevin Wheeler kept working in music for years. In addition to producing bands, Kevin pretended he had three different record labels. He would send recordings and press releases to music publications, trying to get them to write about the bands. If any of them bit, he figured he could always pay for someone to put out a real record. That gambit didn't pay off. In 1999, he and a cDc friend moved to New York to make it bigger in music, but it was tough slogging. They did two off-Broadway soundtracks before Kevin's partner met a woman and moved to Taipei with her in 2001. Then the September 11 attacks wiped out the office where Kevin worked and he got laid off. The biggest problem was his partner leaving, because Kevin always did better as part of a team. "I'm not an Oxblood," he said. "I was the front man. I'll help promote it, spin it, make it funny. I'm at my most productive when I have a partner pushing me, like Gibe in the beginning."

One other music partnership did produce a minor hit. While in high school, a budding writer named Hugh Gallagher, frustrated with unrealistic expectations for college entrance essays, wrote a wonderful spoof with cDc-style self-mocking grandiosity. It concluded: "I breed prizewinning clams. I have won bullfights in San Juan, cliff-diving competitions in Sri Lanka, and spelling bees at the Kremlin. I have played Hamlet, I have performed open-heart surgery, and I have spoken with Elvis. But I have not yet gone to college." Besides getting Gallagher into New York University, it won him a national contest, modest fame, and writing assignments for *Rolling Stone*. Later, he created the character of a Belgian rapper, Von Von Von, and gave a televised performance of a song at Harlem's Apollo Theater with music Kevin wrote and produced. Gallagher gave him a shout-out by name from the stage in a video seen more than a million times on YouTube. After flirting with a career playing poker, Kevin turned to trading currency from his apartment. He reverted to being the shy person he really was when not writing under a handle or promoting a cause or a colleague with onstage antics.

By 2018, the Cult of the Dead Cow had likewise faded into the background. The Def Con and HOPE performances had been over for years, and many hackers in their twenties had not heard of it, unless they picked it up from researching the past or heard about the group from older friends and mentors. Almost everyone had heard about hacktivism, even much of the general public, but they usually associated it with Anonymous or other cDc successors. As with any great teacher of other teachers, the most obvious legacy of the group was the actions of those who were inspired by cDc and the next generation that they reached. That included a large swath of nonprofit activists and researchers and some of the top security minds in government and industry. At Google, security team founder Heather Adkins had grown up on Internet Relay Chat channels with cDc and taken its disclosure lessons to heart. They had set the foundation, she said, for efforts like Project Zero, Google's wide-ranging team for finding bugs in any software and setting a three-month calendar for public revelation. In four years, the group found 1,400 vulnerabilities and drove ninety-day turnarounds from 25 percent to 98 percent. "Disclosure is more mature now, but the stakes are a lot higher," Adkins said. "Companies have a responsibility to protect users. How do you show them the way? Historically, tech companies had only one goal, and that's to make money. Not until someone comes in and disrupts that does it have an impact."

Those individual cDc members who were still accomplishing great things on their own, fulfilling some of the group mission, included Oxblood, Mudge, Christien, and one more person: Psychedelic Warlord. Like the others, he saw that as technology became more central to life, the critical thinking that had grown up with it likewise needed a bigger platform.

> CHAPTER 13

> # THE CONGRESSMAN AND THE TROLLS

WHEN SHE LEFT the Cult of the Dead Cow mailing list in 2006, Carrie Campbell's farewell note included a tribute to the man who had brought her into the tribe in the 1980s. "By the way, Psychedelic Warlord is moving up rapidly in politics in El Paso. I'm so proud of him. We seriously need to not claim him so his career can progress unhindered." She included a link to the Wikipedia page about the Texan under his real name: Robert "Beto" O'Rourke.

Some newer members had not known his name, and most had never met him. Beto's last outing with the group had been the reunion at the HOPE conference in 1997, the same year that Laird Brown floated a story to the audience about the Hong Kong Blondes. But everyone in cDc honored Carrie's parting wish. His youthful participation in cDc remained secret as Beto's profile rose in El Paso. He had gone to Columbia University after his Virginia boarding school, then worked at a New York internet service provider. He also played with a punk band, Foss, that Carrie housed when it came through Seattle. Back home, he started a modest web design business that moonlighted as an alternative news site. Then Beto followed his father into politics, winning a spot on the city council. El Paso was one of the poorest cities in America, and it sat across the river from the mass drug murders in Ciudad Juarez. Beto advocated for liberalized drug laws and wrote a slim book with a city council ally arguing that marijuana legalization would cut down on the gangster profits that were fueling so much bloodshed.

Beto had his eye on a seat in Congress, but party officials advised him to wait until the established Democratic incumbent retired. Instead, Beto took a calculated risk and challenged him in the primary. The veteran underestimated Beto, who outworked him, knocking on sixteen thousand doors. He showed voters the energy he could devote to their interests. He won the primary and the general election as well, joining Congress in 2013.

In 2016, while Beto and others were holding a sit-in at the House of Representatives to force a floor debate on gun control, the Republican House speaker called a recess. That invoked the congressional rule that C-SPAN can't broadcast when the chamber is not in session. So Beto began broadcasting the event from his phone over Facebook, and the network aired it. The stunt drew wider attention to the majority party's refusal to even deliberate on a vital issue, and it showed Beto's willingness to think like a hacker to work around the established technological, political, and media procedures.

As a Democrat in majority-Republican Texas, Beto usually handled minority status with grace. After a blizzard canceled flights from Texas back to Washington in March 2017, he embarked on a twenty-nine-hour road trip with the Republican from the next district over, the equally pragmatic former CIA operative Will Hurd. As they drove back to Capitol Hill, they streamed live video as they chatted, answered questions from viewers, and listened to music. They talked about Russian interference in the election, the proposed border wall, and health-care legislation as they got to know each better. The video went viral, garnering millions of views.

After Trump's election, Beto knew there wasn't much he could get done from the House. Even if the Democrats won a majority, it would take many years for him to move up enough in seniority to lead a major committee. On the other hand, if he managed to pull an upset again, this time in a Senate race against Texas Republican Ted Cruz, he could help flip that body and start making a difference right away. Under the law, Beto would have to first surrender his House seat, since he could not run for both reelection to that office and election to the Senate. It would be all or nothing.

When Beto declared for Senate in early 2017, Republicans controlled the White House and both houses of Congress, Texas had not elected a Democrat in a statewide election in decades, and Cruz was among the best-funded members of the Senate. Cruz had added to his prominence by being the last plausible Republican standing in the 2016 primary before Trump beat him out. And Trump was off to a strong start in the White House, with James Comey still leading the FBI and no special counsel investigating possible collusion between Russians and the Trump campaign. Beto, meanwhile, was more liberal than the average Texas Democrat, making him an easy target for Cruz's many taunts.

But Beto had advantages as well. Cruz had high negatives in polling alongside high positives. Trump's support had slipped in public polls, which hurt all Republicans. Beto, meanwhile, had communication, community-building, and critical-thinking skills that dated to his days as a fledgling hacker. If he believed that a popular policy was wrong, he said so. Beto's technological savvy, while not in the same league as Mudge's or Christien Rioux's, put him way ahead of the average member of Congress on the subject and helped him appeal to younger voters as well as those increasingly concerned about tech threatening privacy and traditional jobs while spreading falsehoods. Certainly, he was a sharp contrast to those members of Congress who questioned Facebook chief executive Mark Zuckerberg and stumbled over such basic concepts as its advertising-dependent business model and how Facebook differed from Twitter.

Beto's familiarity with tech also helped him reach funders in Silicon Valley and elsewhere. cDc members quietly whispered about his history to a few of the most trustworthy and wealthy tech people they knew. A friend hosted an early fundraiser in Los Angeles, and Sam Anthony held his in Boston. Carrie had an emotional reunion with Beto at one in Seattle. Beto told everyone he had stayed in Carrie's house as a

punk-band bassist and had eaten all her Cheerios. As the event wound
down and she thought of all Beto had accomplished in the years since
she had last seen him, Carrie teared up with pride as she hugged her
old friend goodbye.

[x x]

It wasn't just that politicians needed to think more about technology
and its unique multidisciplinary role in the world. Those in technology
needed to think a lot more about politics. Trump's election sparked a
desire in many to fight back against what they saw as domestic information warfare. Security experts felt a special twinge, because hacking
into the Democratic National Committee, the Democratic Congressional Campaign Committee, and John Podesta's Gmail account had
played a pivotal role in the election. Following the model that Jake Appelbaum had helped promote, the contents of those emails were spread
by WikiLeaks, reported by the partisan and mainstream press, and
circulated wildly over social media.

Over the course of 2017, as evidence emerged of the depth and sophistication of the efforts to promote division and Trump on Facebook,
Twitter, and Instagram, a wider swath of the American public turned
against tech companies. Inside the Silicon Valley giants, divisions crystallized. A minority were unapologetic Trump supporters, like Palantir
cofounder and Facebook board member Peter Thiel, or took his ascen-
dancy as an opportunity to speak out against what they saw as discrimination against straight white men, like a Google engineer who claimed
he was fired for writing about internal bias.

But many more felt caught up in a moral crisis unlike any the Valley had ever faced. Some wanted to use the money that they had earned,
their networks, and some of their tech skills to set things right. Start-up
founder Maciej Ceglowski was watching as the president issued a ban
on immigration from multiple Muslim-majority countries. Such policies grated especially on many in tech because an outsize number of
founders and employees came from other countries. And some who had

excused the expansion of tech surveillance under previous administrations now fretted that such powers were in the hands of an executive branch that was openly disdainful of judicial review.

Ceglowski began holding meetings of concerned employees under the banner of Tech Solidarity. One offshoot from those meetings, led by Slack engineer and Jake Appelbaum victim Leigh Honeywell, created a public "Never Again" pledge to oppose immoral conduct and go public if necessary, which has been signed by more than 2,800 employees. Among other things, the signatories promised to advocate against retaining data that could be used for ethnic or religious targeting and advocate for deploying end-to-end encryption.

The Solidarity meetings raised money for immigrants' lawyers and coordinated volunteer coding projects. As the 2018 midterms approached, confronted with billionaires on the other end of the spectrum spending untraceable "dark money" to push right-wing candidates, Ceglowski fought back with what he called "dork money," funding a slate of progressive candidates around the country in districts he thought he could flip. Among his small circle of coconspirators was Adam O'Donnell. They also advised dozens of campaigns on security, hoping to stave off a repeat of the devastating 2016 hacks.

cDc itself contained almost no Trump supporters. But because of its multifaceted legacy, it had protégés on both sides of the fight. They battled on Facebook's pages and even inside that company's executive ranks. On the right were some especially voluble members of cDc's farm team and fan club, the cDc Ninja Strike Force.

Rob Beck, cDc's friend from Microsoft and @stake, had been in charge of NSF for a while, and then others took it on. Membership got looser, group founder Sam Anthony said, and "one branch became this awful Gamergate, neo-Nazi and Russian intelligence nexus that is ruining the world." Organized on 4chan and other sites, Gamergate's organized trolls went after female gaming journalists with mob attacks on social media before eventually coalescing behind Trump. By 2012, the NSF mostly lived on as a Facebook group. Members posted links to security advisories, breaches in the news, and whatever else they found interesting. Some of the members, though, were 4chan veterans who wanted to provoke, and they resorted to posting racist cartoons and jokes. Several considered it harmless trolling and denied being racist. But many core cDc members were deeply offended. "All these people were influenced by cDc. But there was no structure or indoctrination or social vetting," said Beck, who reconnected to NSF after years away and found it jarring. Beck began sending some of the more extreme posts to Sam, Luke Benfey, and Kevin Wheeler, just to make sure they were aware.

In June 2012, Luke wrote to the cDc email list with links to racist caricatures from the NSF Facebook page. "I think it is deeply, deeply shameful that this sort of thing is being associated with cDc," Luke wrote. Paul Leonard, who had maintained relationships with some of the offenders, agreed that the drift in NSF had gone too far. "Some of the guys are right wing, some are just 'edge lords'—they don't have any particular ideology, they just like blowing things up." Paul wrote to two of the offending posters: "I generally consider the NSF guys to be people I want to associate with, and [for] most of the NSF this really is the case. Now I have to question this relationship and it pisses me off. I don't really care about your politics, I don't even care about your racial beliefs whether they're heart-felt or just trolling. It bothers me that you don't seem to have any kind of internal editor that can differentiate between amusing offensiveness, and the kind of tedious, boring, lowest common denominator type of offensive material that shouldn't even make the grade."

Laird Brown wrote to Luke: "This is a painful thing to write. If this situation continues and nothing is done about it, I would be forced to withdraw from the cDc and take Hacktivismo with me. I can't be associated with this bile." They had limited options, since NSF hacker Colton Sumners had founded the Facebook group in 2007 and kept administrative control. "I gave all the misfit toys a voice," Sumners said. Luke reported the NSF posts to Facebook as offensive. As the group contemplated more drastic action, Kevin reappeared and got caught up. Then he wrote to Sumners: "This is worse than I thought. I need admin access to the NSF group, and the racial stuff has to go. Bottom line, the stuff isn't strategic and it's causing problems for me. Thanks!" After a protracted struggle among the admins, the old cDc guard wrested control back.

Sumners, Xerobank Tor browser creator Steve Topletz, and a few others in NSF, including self-described black hat and white nationalist Timothy "Matlock" Noonan, had their own publication and social group, DSSK Corp. Speaking as Matlock, Noonan declined to discuss most of his activities with me. He did say he had grown up on cDc but that the crew had become stale and irrelevant and didn't do much hacking. He admitted to one illegal hack on the record, a 2012 takedown of a pedophile site targeted by Anonymous. cDc legend Chris Tucker, known as Nightstalker, had recently died, and Noonan published a press release under the cDc and NSF banners claiming the attack was Nightstalker's. Noonan and Topletz had also done favors for the US government, including turning over traffic they had found after hacking Iranian servers. Like the old cDc text files, DSSK chronicled various adventures. But the politics were very different. In 2015, a DSSK article reported on a trip by Noonan to Eastern Europe to see Andrew Auernheimer, known online as weev, perhaps the most famous troll of all time. Paul Leonard had known Auernheimer from childhood, back when he wasn't racist. "He was mostly an irritating edge loser who played with racism for fun and lulz, up until he went to jail," Paul said, echoing others. "When he got out, he was notably changed."

By the time of the DSSK visit, Auernheimer had a swastika tattoo and was staying in countries that didn't extradite people to the US. He lived in Ukraine for a time, and then a Russian breakaway republic in Moldova. He handled the technology behind the Daily Stormer, a Nazi and pro-Trump site run by Andrew Anglin, who spent enough time in Russia proper to send in an absentee ballot from there. Both men exhorted racist protestors to turn out in force at the Charlottesville, Virginia, march where a Stormer fan drove into and killed counterprotestor Heather Heyer. It is not yet clear what else Auernheimer got up to in helping Trump. But Auernheimer was suspected of hosting faked documents in the French election of 2017 to help the far-right candidate Marine Le Pen.

The rot in NSF spurred cDc proper to go further to make up for the Frankenstein's monster it had created.

[✕ ✕]

While cDc and a transformed NSF battled on Facebook's pages, a more momentous conflict was brewing behind the scenes at Facebook the company, arguably ground zero for the election misinformation battle. Beyond overt support from Thiel, who spoke for Trump at the Republican National Convention, Facebook's collection of data on its users, as well as its lax policies about what apps could collect from whom, allowed a secretive network of companies including Cambridge Analytica to collect material on as many as 87 million Americans. The companies, funded by billionaire Republican donors Robert and Rebekah Mercer, claimed they could tell from the psychological elements of that data which ads would be most effective to show to whom. Famously, the information went to help Trump. But before that, during the primaries, it was helping rival Republican presidential candidate Ted Cruz, the Texas senator Beto would face two years later.

Also inside Facebook was Alex Stamos, the former consultant at Mudge and Christien Rioux's @stake. After Snowden had revealed that many tech companies worked closely with the NSA, then iSec Partners cofounder Stamos had given a talk at Def Con 2013, arguing that security experts had broader social responsibility and should consider quitting before harming the public. "I'm a corporate white-hat sellout," Stamos admitted up front. "This talk is about how, if you decide to be a corporate white-hat sellout, you can do that as ethically as possible." Stamos described how ordinary companies were now getting sucked into cyberwarfare, citing as an example an advanced cryptographic attack on Microsoft that allowed for the overseas installation of malware attributed to the US, as well as constant attacks by the Chinese on Google and other tech companies. As the grandson of poor immigrants,

Stamos said that he was proud to be an American, but that his loyalty belonged more to the values of the country than to any temporary set of leaders. Like doctors, he said, technologists and especially security practitioners held critical roles that might require moral obligation. "Maybe this means all people deserve for their technology to be trustworthy," he suggested, before posing a series of hypotheticals and asking for a show of hands to see who would do what.

In the first scenario Stamos gave, you discover a major flaw. Do you announce it, sell it to your government, sell it to the highest bidder, use it yourself, use it as leverage to get a consulting deal with the software maker, or work with the vendor to patch it and then disclose? Most picked the last option, the coordinated disclosure pioneered by the L0pht. In another case, what if national-security authorities want to have an informal chat? Do you accept the meeting, avoid it, or ask for an email to run by your company's lawyer? About a quarter of the audience members said they would meet even after Snowden. Stamos said he had until recently felt the same way, and he had spent hours talking to the authorities a year ago at the same conference. Now, he said, he would involve a lawyer. What about finding a corporate back door collecting data from your customers, if your boss says forget about it? Do you ignore it, escalate it within the organization, quietly look for a job elsewhere, or publicly quit and break your nondisclosure agreement with the company by explaining why? Stamos said he was between the last two answers.

He closed by urging the older and more experienced professionals to share the tough decisions they had made with newer entrants, and all to think about scenarios to come so they wouldn't be caught off guard. "Try to live an examined life," he said. Later, in 2013, I revealed that security company RSA had taken $10 million to put an NSA back door in a tool kit it distributed for protecting websites. Stamos was one of ten speakers who as a result pulled out of the early 2014 RSA Conference, the only US security-industry gathering larger than Def Con, and one founded, ironically, to contest federal demands for control over encryption. Instead of staying home, Stamos organized a counterconference,

TrustyCon, held at the same time as the RSA Conference but devoted to examining how to respond to improper government influence and other threats to security and privacy.

For all his outspokenness, Yahoo hired Stamos as chief information security officer later that year, partly in response to the government tricks exposed by Snowden. But Stamos quit Yahoo in 2015. He told his staff that he left because new Yahoo CEO Marissa Mayer had not alerted him about or challenged an order from the US Foreign Intelligence Surveillance Court, which meets in secret to approve wiretaps on suspected international spies. This order required Yahoo to install new software to scan every single email processed by its system for a certain digital signature, such as a cookie installed on a single user's computer or an encryption key. After Stamos's team found what seemed like a hacker's rootkit installed on Yahoo's email servers, they feared that Russian hackers who had bedeviled them in the past were back, and they alerted Stamos, who called everyone into the office at 5 a.m. on a Monday. When email engineers suggested he check with the legal department, Stamos did, learned that the tap had been authorized, went up the chain to Mayer, and quietly departed.

The US intelligence agencies' target had been legitimate. But the means by which they sought the suspect's correspondence, and the undisclosed complicity of Yahoo, made a mockery out of the annual transparency reports in which Yahoo estimated how many email accounts it had examined for the government. And if it would search every last email for one secret piece of information, what would stop it from doing it again for a phrase, like one expressing hostility against the current president, or the next one? This story led some to proclaim that Stamos had set himself up as a sort of human canary in the coal mine, and that if he ever left Facebook, users should take heed.

[x x]

At Facebook, Stamos primarily protected the company itself. He also stood between users and organized criminals, sexual predators, and fraudsters. But during the 2016 election, he was on alert for activity by a group known as APT28 that was linked to the GRU, Russian military intelligence. It was one of the groups that hacked the Democratic National Committee, and Stamos's team found it was behind a Facebook page dedicated to DCLeaks, a short-lived spot for pushing out hacked Democratic emails. They ruled that DCLeaks was "inauthentic"—a politically neutral grounds for banning—in August 2016, but internal worries about appearing partisan delayed the ban until October. Stamos's side won the fight only after DCLeaks posted phone numbers associated with financier and Russian nemesis George Soros, violating a Facebook policy on exposing personal information.

After the election and before Trump took office, US intelligence agencies unanimously agreed that Russia had interfered to help Trump win and that it had spread fake news on Facebook in the effort. An internal Facebook task force looked into it and found mainly money-driven spammers trying to get people to their pages with slanted stories. Bigger priorities at Facebook were the coming elections in Western Europe, where France and Germany pressed for help. Working with France's defense-only cybersecurity agency, ANSI, Facebook experts found GRU reconnaissance of campaign workers and tens of thousands of fake French accounts connected to Russia that amplified divisive stories. Only after US intelligence officials told *Time* magazine that Russian propagandists bought Facebook ads did the company realize that ads were a vector, and one with lots more forensic data.

Stamos's team dove in and found a massive cluster from the Saint Petersburg–based Internet Research Agency, and for the first time they saw that the fake activists were pushing far-left narratives and memes as well as far-right ones. A closed-door briefing to Special Counsel Mueller triggered legal requests for content that formed the basis of his 2018 indictments of thirteen Russians and three companies for interfering in the US election. The most important of those companies, Russia's IRA, had bought thousands of ads stoking divisiveness on social media. One on Instagram showed black men attacking a police offer; the ad was shown to users who had expressed interest in either Fox

News or Senator Ted Cruz. Cruz also got tweets of support from IRA accounts.

Stamos was trying to do the right thing, but it came at a great cost. Executives above him repeatedly minimized the Russian activity in his public reports. When he briefed Facebook's board about what he had found in September 2017, the directors asked him if he had successfully rooted out all of the stealth accounts. He answered, truthfully, that he had not. The board members then grilled CEO Mark Zuckerberg and chief operating officer Sheryl Sandberg about why they hadn't told them how bad it was. Sandberg paid the tongue-lashing forward, yelling at Stamos: "You threw us under the bus!"

Stamos never controlled all of the security apparatus at the company, and the board flare-up cemented his reputation for being overly aggressive. In December, when Stamos suggested reporting to someone besides the general counsel, other executives in charge of Facebook's main service and engineering stepped up and said they could handle security interpreted more broadly, now that it was a subject of global concern.

Stamos, who had increased his staff from 60 in 2015 to 120, was boxed out and left with three employees and a vague mandate to do something about election security, like advising on plans to fight propaganda. He pushed for wide collaboration in that, and he and Google's Heather Adkins helped the Defending Digital Democracy project at Harvard's Kennedy School of Government, which in turn advised campaigns and trained election officials from thirty-eight states.

In the scope of professional ethics he had laid out at Def Con in 2013, Stamos had followed through. But his control was smaller than many outside the company had understood, and he could have made that more widely known. Then again, he became chief security officer at Facebook before anyone knew that the most important battle was going to be propaganda, not electronic breaking and entering. Stamos negotiated to leave Facebook in August 2018 for a post at Stanford University, an attempt to set up a forum there for the big internet companies to thrash through tough issues on neutral ground, and a planned book on fighting information operations. "I'm not a big company guy," Stamos said. "Above my level, it's *Game of Thrones*." In a farewell memo, he wrote that he accepted a share of blame for the election manipulation and urged those remaining to "deprioritize short-term growth" and "be willing to pick sides when there are clear moral or humanitarian issues."

When the *Guardian* broke the news that secretive government and political consulting firm Cambridge Analytica had obtained the personal information of tens of millions of Facebook users through a misleading quiz and that it had not been deleted when Facebook asked, Stamos couldn't really be made to take the fall. Technically, it was not a breach: it was a failure of basic advertising processes, where Stamos had no control. Instead, CEO Zuckerberg had to go before Congress to apologize and promise to give users more control over their data. Meanwhile, former Facebook and Google executives began condemning their former employers for allowing disinformation to thrive. Tristan Harris, a former design ethicist at Google, formed the Center for Humane Technology and warned that Facebook and YouTube had let some of the world's most powerful instances of artificial intelligence figure out how to keep people watching, and that the answer had been to show outrage and extremism. Rank-and-file tech workers forced Google to drop a contract to provide artificial intelligence for analysis of Pentagon drone footage that could be used in automated targeting, and protests spread through Silicon Valley over tech employers' contracts with border authorities who separated children from immigrant parents.

[x x]

In 2017, Adam O'Donnell got hold of Stamos to talk about Beto. O'Donnell knew Stamos from the iSec days and admired his ethics and comfort with politics. He also knew that Stamos would honor cDc's biggest secret, that Beto was one of them. Stamos was as delighted as he was shocked.

"I have to support this guy, someone who has been active in this world since he was a teenager," Stamos said. "You can see people like

Beto and Hurd working together." Stamos told friends that Beto was a savvy nerd, leaving his cDc membership unmentioned. In November, Stamos joined in cohosting the Beto fundraiser at Adam's house. At that event, Beto laid out his beliefs and approach. Not only did Beto share much of cDc's and the broader Silicon Valley community's feelings on tech policy, but the same was true for their feelings on legalizing soft drugs. Like Kroupa and the EFF's founding libertarians, Beto saw the profits from illegal drugs fueling the murders that were ruining Mexico and reaching across the border. (Beto had also sided with Republicans in a number of votes to reduce regulations and taxes.) The easiest thing to tar him with, Beto said, was a pair of arrests from twenty years before, one for jumping a fence and the other for driving after drinking. Neither led to convictions. Beto said he spoke to others with worse records who couldn't vote as a result, and he believed in second chances.

Relentlessly upbeat, Beto's fortunes rose as Trump careened from crisis to crisis, continually intervening to help Russia even when his top aides had publicly promised retribution. Cruz followed Trump despite his earlier protests. "Cruz is a rare and precious gift. He's so loathed that any passable Democrat with a picayune chance of tipping him was bound to draw more attention and inspire more hope than the political dynamics warranted," Frank Bruni wrote in his Sunday *New York Times* column in April 2018, a month after Beto won his contested primary. "But Beto is more than passable. Many of his campaign events are mobbed. People line up for selfies and then insist on hugs." Bruni noted that Beto was fluent in Spanish, classic punk rock, and contemporary country music, and that *Vanity Fair* had dubbed him Kennedyesque. Beto had taken in more money than Cruz by mid-2018, and polls put Beto closer and closer to Cruz, eventually within the margin of error, with his biggest handicaps being the lack of name recognition, which would sort itself out by the fall vote, and the history of low voter turnout, which he was trying to reverse. "People are coming out because they don't want a wall," Beto told National Public Radio. "People are coming out because they don't think the press is the enemy of the people, they think they're the best defense against tyranny." Beto rejected consultants and big data as well as corporate PAC money, taking in small individual donations and going from town to town, running "in the most punk rock way I know."

Beto didn't say as much, of course, but it was also the most hacker way he knew. From speaking his mind in the underground newspaper of his era, the online bulletin boards, and then in the alternative news site in El Paso that got him launched into politics, Beto had learned to seek new ideas, to be comfortable in his own skin, and to speak his mind with humor, connecting honestly with those who differed rather than seeking to conform. He had stuck to that as greater numbers of people also began looking askance at the dominant political structure that Trump had taken to new and uncomfortable places. Cruz, by contrast, looked very much like an opportunist and a creature of that power structure, one who asked Trump, whom he had called a pathological liar, to come campaign for him in Texas.

Traveling to all 254 Texas counties, including those long written off as hopelessly Republican, without any polling or focus groups—that was hacking politics. With the biggest crowds yet coming to see Beto in September 2018 even in deep Republican territory, his small staff used iPhones and social media to share authentic moments that spread widely. In one, a man asked whether he shared his upset at football players kneeling during the national anthem. Trump had repeatedly raised the issue, calling the players, many of them black, unpatriotic "sons of bitches." Beto, who had never been asked the question before, thanked the man for the question, and he thanked the veterans in attendance for their service. Then he gave a short, spirited, off-the-cuff history of peaceful protest over racist policies and violence and how it had helped change the Deep South. "I can think of nothing more American than to peacefully stand up, or take a knee, for your rights, anytime, anywhere, anyplace," Beto said. The video was viewed more than 40 million times and won Beto national television coverage.

As the newspaper charts predicting control of the Senate kept shifting to the left, one showed that if all other races went as projected, control of the body would depend on whether Beto won. Even if he lost, it was hard to imagine Beto fading away. He seemed destined for the national political stage in some way. As the election neared with him still on Cruz's heels, comparisons shifted to Obama's first run, and pundits spoke of a future Beto run at the presidency. "O'Rourke offers not just a path to victory in Texas but an antidote to the entire stupid artifice of American politics in the Trump era," blared *Vanity Fair*. "He's authentic, full of energy, and stripped of consultant-driven sterility. On what planet is Beto O'Rourke not a presidential contender, even if he loses?"

[x x]

Though Beto had been obliged to discuss his 1990s arrests, nobody knew about his teenage hacking, let alone his long association with what was for many years the best-known group of technology-minded miscreants in the world, and understandably he did not volunteer the information. Yet when I told him I wanted to include his background in a post-election book, he was willing to talk. He was unabashed about how much being part of the Cult of the Dead Cow had meant to him.

"It's cool to be connected with the people who were in cDc and people who were involved with early internet culture," Beto told me. "I was really at the margins, but I very much wanted to be as cool as these people, as sophisticated and technologically proficient and aware and smart as they were. I never was, but it meant so much just being able to be a part of something with them."

Born a couple of years after the cDc founders, Beto said he was like the other early members in some respects. "I had a really hard time fitting in and just finding a path along the conventional route," he said. Beto's father, a prominent local judge later killed in a bike accident, brought home an Apple IIe with a 300 baud modem while Beto was in middle school, and he went searching for bulletin boards. Beto found a few in Texas, including Kevin's Demon Roach Underground in Lubbock. The long-distance boards required pilfered dialing codes "so

I wouldn't run up the phone bill," Beto said. Part of the attraction was that boards were "a great way to get cracked games." Later, he realized he had made poor choices on that matter.

But the games were not what made Beto keep coming back. "Being part of that, starting my own bulletin board, all of it was fundamentally just wanting to be part of a community," he said. It was also about a search for culture apart from mainstream movies and the records on the radio, market-tested and inauthentic and boring, at least to Beto and other teens in search of their own identities. "This was the counter-culture: *Maximum Rock & Roll*, buying records by catalog you couldn't find at record stores. cDc was kind of a home for people who were inter-ested in that part of the subculture." Beto was searching for knowledge too, "in terms of understanding how the world worked—literally how it worked, how the phone system worked and how we were all connected to each other. They were all free-thinking people within cDc."

Beto logged in the most during the late 1980s and very early 1990s, before starting at Columbia University in 1991. He checked in at times later, feeling closer to the cultural wing of cDc than the star technolo-gists. Though not an accomplished programmer, after graduating from college Beto built websites and set up high-speed connections in New York. Then it was back to El Paso, and his own small software and services company, and the series of underdog political runs that began with an upset putting him on the city council. Among other unorthodox moves there, Beto sponsored a 2009 measure calling for a national dis-cussion about legalizing marijuana.

Beto credited cDc with influencing his thinking in a number of ways that he had brought to bear already. Not least, he was fighting to restore net neutrality, which prevented internet connection provid-ers from favoring some web content. "I understand the democratizing power of the internet, and how transformative it was for me personally, and how it leveraged the extraordinary intelligence of these people all over the country who were sharing ideas and techniques," Beto said. "When you compromise the ability to treat all that equally, it runs counter to the ethics of the groups we were part of. And factually, you can just see that it will harm small business development and growth. It hampers the ability to share what you are creating, whether it is an essay, a song, a piece of art. And so that experience certainly informs what we're doing here now."

Beto said his history made him want to push a broader discussion about making the most out of gifted technologists and other thinkers with unconventional ideas, which can have more impact because they break with patterns and tradition. That was the same insight Mudge had brought to DARPA—that adding similar but more complicated defenses doesn't help. "There's just this profound value in being able to be apart from the system and look at it critically and have fun while you're doing it," Beto said. "I think of the Cult of the Dead Cow as a great example of that. In doing that, you make our overall society stronger, as with just the vulnerabilities technologically that people were able to uncover and point out and be part of fixing."

"There was something really valuable about the counterculture and countersystem. Putting those talents to use, you make things better for everyone, and it should not lock you outside of security, or being productive, or taking a more conventional path. I'm an example of that, from starting a business with friends, to city council, Congress, and running for senator. Part of my success was being exposed to people who thought differently and explored how things work. There are alternate paths to service and success, and it's important to be mindful of that."

On Election Day 2018, Beto drew the votes of 4.02 million Texans to Cruz's 4.24 million, losing by less than 3 percentage points. Turnout surged in Democratic areas, with a half-million more people voting in the Houston area alone than had in the previous midterm. That enthusiasm swept all Republicans out of power in the state's biggest city, flipped two House seats, and took a larger chunk of the state legislature. University of Texas government professor James Henson declared it "the beginning of the end for one-party rule in Texas."

It was part of a broader rebuke of Trump that won Democrats the House and their first meaningful check on the president's power. While upset about Beto's loss, Democratic leaders and activists were so thrilled with his performance in the state and on the wider national stage that the debate turned to whether he should run for Texas's other Senate seat in 2020 or instead seek the presidency, since the country as a whole was significantly less conservative than Texas. On the front page of the *New York Times*' first Sunday edition after the election, the paper's political reporters wrote that Democrats were debating whether to run in 2020 as moderates or as liberals, and that "at the center of the dispute is Rep. Beto O'Rourke," in the latter camp. A month later, as Beto looked increasingly likely to run for president, the paper wrote that he had realized a long track record mattered less in the age of Trump than grassroots enthusiasm. As Beto mulled his next step, the hidden hacking history he knew would be revealed in this book played a role.

As with his youthful arrests and brief punk rock career, Republicans would certainly use his teen writings and his associations to tar him as a cultural misfit and radical. But the legacy would also engender deeper loyalty from some Silicon Valley technologists who were already drawn to his appreciation of their issues, his liberal approach on some matters, and his libertarianism on others. Forced by the 2016 presidential election to consider their own role in society in a less flattering light, those technologists could see Beto as a powerful shot at redemption.

Beto finished his last term in Congress and spent most of January driving himself through several states talking to whomever he encountered, weighing what to do next. When he returned, he booked an interview on Oprah Winfrey's show *Super Soul Sunday* and teased the political class and the public, saying he would decide soon. Just after the release of an upbeat documentary about his Senate run and a *Vanity Fair* cover story on his background, Beto declared his campaign for president on March 14. "This is a defining moment of truth for this country," he began in his announcement video. "The interconnected crises in our economy, our democracy, and our climate have never been greater, and they will either consume us, or they will afford us the greatest opportunity to unleash the genius of the United States of America."

The day after his announcement, with Beto immediately in the up-per echelon of the race and our embargo on his youthful activity ex-pired, Reuters released a long story based on this manuscript. The news shot around the country, making every major paper and website and creating more than 50 million impressions on social media. On Fox News, Tucker Carlson mocked the vulgar poem to a cow Beto had posted as a teen. On HBO, Democratic show host Bill Maher gave a different perspective in his opening monologue: "Some interesting par-allels between Trump and Beto. As a teenager—this is true—Beto be-longed to a computer hacking group called the Cult of the Dead Cow. And as an adult Trump belongs to a computer hacking group called Russia."

The overall impact appeared modestly positive for his nascent cam-paign. Many younger voters who had been unimpressed by Beto before said it was the first thing that intrigued them, and technologists were positively giddy. "This is one of the most 'holy shit' things that has hap-pened on my twitter feed that wasn't bad...ever," tweeted Robyn Greene, a Facebook privacy manager formerly at the ACLU. When asked about it, Beto said he was "mortified" at some of his early writing. Looking on nervously, the core of the cDc hoped they were right that the disclo-sure would nonetheless help him appeal to the young and disaffected without losing too many traditionalists. Beyond that, they hoped Beto would open up about his roots and, win or lose, show that a future run by hackers could be a great one.

Robert "Beto" O'Rourke / Psychedelic Warlord

>   **EPILOGUE**

IN ITS EARLIEST days, the chief moral issues for the teens in the Cult of the Dead Cow were how badly to abuse long-distance calling cards and how offensive their online posts should be. But as they matured, the hackers quickly became critical thinkers in an era when that skill was in short supply. In an evolution that mirrored and then led the development of internet security, cDc went on to forge rough consensus on the complex but vital issue of vulnerability disclosure, to show that enabling strong security could be a viable business, and to merge the hacking spirit with activism on behalf of human rights. It also kept a remarkably big tent, roomy enough to include support for acts of civil disobedience as well as work for the military, as long as both were principled. They all helped push a realistic understanding of security challenges and ethical considerations into mainstream conversations in Silicon Valley and Washington. As the big picture in security grows darker, those conversations are the best hope we have.

One lesson from the Cult of the Dead Cow's remarkable story is that those who develop a personal ethical code and stick to it in unfamiliar places can accomplish amazing things. Another is that small groups with shared values can do even more, especially when they are otherwise diverse in their occupations, backgrounds, and perspectives. In the early days of a major change, cross sections of pioneers can have an outsize impact on its trajectory. After that, great work can be done within governments and big companies. Other tasks critical for human progress need to be done elsewhere, including small and mission-driven companies, universities, and nonprofits. It gets harder to keep the band together over time, but cDc's impact lives on in those whom members hired, taught, and inspired. That said, a movement cannot control its children. The Citizen Lab and Tor are one thing, while Lulz Security and Gamma Group are another. Trolling and fake news also owe something to cDc, and neither is anything to be proud of.

As I was nearing the end of the writing process, a moderately well-known security professional asked his Twitter followers for some current ethical issues facing the industry. His feed was inundated with questions. If you live where encryption is outlawed, do you help activists encrypt anyway? If you discover a malware campaign that appears aimed at a reviled terrorist group, do you expose it? If you make a monitoring tool, do you sell to nonsanctioned but repressive regimes? If authorities want you to sell a zero-day vulnerability to a broker instead of warning the vendor, do you? If your government asks your antivirus company to search on computers for a specific signature that is not malware, do you? The questions will go on forever, and there need to be better ways of getting debate and answers. One thing that would help is a shift toward public-interest technology like that of the Citizen Lab. Lawyers are expected to do charity work, and there are plenty of public-interest jobs, noted author Bruce Schneier. Neither is true for technologists yet.

Beginning around 2000, after most of the people in this book had left college, accredited US engineering and computer science programs were obliged to require some education in ethics, typically a single course. Too often, those courses are taught by philosophers with no grounding in practical work. The best texts in the field use case histories, such as the *Challenger* space shuttle explosion. Before that disaster, an outside engineer on the shuttle had recommended against a cold-weather launch. He then let his management talk him into changing his mind.

Some of the top professional associations, such as the Institute of Electrical and Electronics Engineers, have slowly evolving ethical codes. But their membership is limited, the codes are enforced only if someone complains, and some guidelines are too abbreviated to be of much use when members seek advice. There is no regulation or continuing-education requirement, both of which govern practicing lawyers. Even the canon of security literature isn't that widely read. "Engineers have a profound impact on society," said former IEEE president and current engineering college dean Moshe Kam. "But quite frankly, there is no glory in dealing with this."

Even those who spend considerable energy wrestling with such issues rarely speak in public about it, which means others don't get to learn from them. Facebook's Alex Stamos is one exception. Another is Dug Song, the Michigan security expert who came up in the hacking group w00w00 and founded Duo Security, bought by Cisco in 2018 for more than $2 billion. In a 2016 speech to students at the University of Michigan, Song argued that moral reasoning was fundamental to what should be a noble endeavor, since technology is the only thing that increases human productivity. "Security is about how you configure power, who has access to what. That is political," Song said.

Rather than thinking about the world as binary, good or evil, Song said he found it helpful to think of the matrix in the role-playing game Dungeons & Dragons, with one axis running from good to evil and another one running from lawful to chaotic. Darth Vader, he explained, is lawful evil: he wants order, it's just for a bad cause. In that vein, he described w00w00 as neutral on both axes. On balance, Snowden might have been chaotic good, and the NSA might have been lawful evil, he said. *Phrack* was chaotic evil, L0pht lawful neutral, and, Song told me, cDc was chaotic good. Whatever the law says, Song believes that professional ethics requires him to contribute to the social good.

Of all those involved in the burgeoning technology industry, which now includes the world's six most valuable companies, security experts like those in cDc were the first to grapple daily with matters of conscience and immense impact on safety, privacy, and surveillance. But such broad issues are now spreading throughout the tech world. Facebook, Twitter, and YouTube are doing poor jobs of stopping propaganda and are letting automation promote content that is engaging because it is extreme. Google is mulling bringing censored search back to China, which it left on principle in 2010. Yet it bowed to employee pressure and walked away from a Pentagon contract to help analyze drone footage that could be used in targeted killing. Apple fought the FBI on back doors but agreed to store user data in China. Workers at Amazon are protesting that company's sale of facial-recognition technology to police, and those at Microsoft are fighting deals with the Trump immigration authorities that are separating families at the border. Technology as a whole is engulfed in what may prove to be a permanent moral crisis, and the best place to turn for wisdom on how to handle it is the people who have been through this before, whether they serve in giant companies or start-ups, nonprofits or Congress.

The more powerful machines become, the sharper human ethics have to be. If the combination of mindless, profit-seeking algorithms, dedicated geopolitical adversaries, and corrupt US opportunists over the past few years has taught us anything, it is that serious applied thinking is a form of critical infrastructure. The best hackers are masters of applied thinking, and we cannot afford to ignore them.

Likewise, they should not ignore us. We need more good in the world. If it can't be lawful, then let it be chaotic.

San Francisco–Boston–New York–
Washington–Austin–Los Angeles

> # ACKNOWLEDGMENTS

Twenty years ago, some people complained that the Cult of the Dead Cow was seeking too much media attention. I have dealt with this complaint in the body of the book, and I think it clearly gives the group short shrift. On a personal level, I can tell you that not all in cDc were clamoring for the in-depth attention I have given it here.

Some members were willing to help a lot, providing personal information even if it could hurt them, and I want to thank them most of all. Extra thanks to those who let me be the first to identify them as cDc members by their real names: Kemal Akman, Sam Anthony, Luke Benfey, Bill Brown, Carolin Campbell, Matt Kelly, Misha Kubecka, Glenn Kurtzrock, Paul Leonard, Dan MacMillan, Adam O'Donnell, Beto O'Rourke, Charlie Rhodes, Mike Seery, Dylan Shea, and Kevin Wheeler. It is also worth noting that some were reluctant to speak at all. For months, founder Kevin Wheeler would not return messages from Luke, his effective number two for decades, about helping with this project. Only after Luke threatened to send him a singing telegram did Kevin finally agree to discuss his potential participation. I am grateful that he and others came around.

More broadly, most of the people named in this book and many who are not devoted their time and candor, and I greatly appreciate the education. For kindly housing and looking after me during my research trips, I would like to thank Ralph and Shan Logan, Andrea Shallcross and Jonathan Burn, Rachel Layne and John Mulrooney, Barbara Bestor and Tom Stern, and assorted relatives. I am also indebted to a number of talented and hardworking authors who brought clarity to various aspects of historic and current issues in security touched on here, including John Markoff, Phil Lapsley, Fred Kaplan, Ronald Deibert, Shane

Harris, Andy Greenberg, Bruce Sterling, Steven Levy, and Gabriella
Coleman. For those interested in learning more about the bulletin
board era, I strongly recommend Jason Scott Sadofsky's multipart documentary and his text file collection, both publicly available. I would
especially like to thank my keen-eyed editor, Colleen Lawrie, agent
David Patterson, and media advisor Elinor Mills.

I have been fortunate to work since 2012 at Reuters, which has
some of the finest journalists in the world. The company provided me
the opportunity to pursue challenging and at times risky stories that
paved the way for this book. Beyond that, Reuters graciously gave me
two leaves: in 2014 for my medical recovery and again in 2017–2018 to
report and write the bulk of this book. Good journalism matters greatly,
and I am heartened that more people are supporting it.

NOTES

Chapter 1: An Evening in San Francisco

1   "On a Tuesday evening": I attended this fundraising event; quotes and characterizations are from my notes.

1   "Adam wasn't accustomed to entertaining people": If I cite someone's thoughts in this book, I almost certainly got them from that person in a direct interview. I will note when that is not the case. When I cite some-one's actions, it was because I observed them, was told about them by that person later, or, in a few cases, was told about them by multiple witnesses.

2   "they invented the term *hacktivism*": A cDc critic using the handle Jer-icho has written that the word first appeared in an obscure Minnesota print publication, "InfoNation," in 1995, https://jerichoattrition.wordpress .com/2014/02/17/on-the-origins-of-the-term-hacktivism/. But the dense art review in question uses the word to mean the creation and use of alter-native media, not technological support for human rights. Internal emails from cDc's later Def Con preparations show group members believed they had a new word and worked together to drop it in interviews to push it to-ward common usage.

3   "dating to the group's founding in 1984": This is the beginning year that the founder now gives, but that precedes its first electronic files. A hard-copy cDc zine from 1988 declares the group began in 1986.

4   "Stamos gave a heartfelt talk on ethics": I attended the speech. All of the conference presentations I cite I either witnessed or watched recordings of. The majority are available on YouTube or other sites, but I am not giv-ing web addresses for most of them because they come and go.

Chapter 2: Texas T-Files

9   "Like many of the internet's earliest adopters": The account of Kevin's youth is primarily from Kevin himself. The same pattern holds true for most of the other principals in the book. The majority of the information comes from in-person interviews with the major figures, supplemented by phone or electronic communication.

10   "We have to make our own and truly be elite": This is Kevin's recollection
      of what he said then. More generally, when I quote someone, in the vast
      majority of cases the person quoted spoke those words directly to me, usu-
      ally in person. Sometimes the comments were by phone, email, or other
      electronic messages. If I came by the comment some other way, I will say
      so in these notes.

11   "Gerbil Feed Bomb": Swamp Rat, "Gerbil Feed Bomb," 1985, www.cult
      deadcow.com/cDc_files/cDc-0001.html. Most of the text files I cite are
      still available online via www.cultdeadcow.com or Jason Scott Sadofsky's
      www.textfiles.com. The inclusion of a link here, however, is no guarantee
      it will still be online at publication or thereafter. I will also note that not
      everything on the cDc site is accurate.

12   "KGB 'had some nutty retardo sex & violence stuff'": This is from an
      email to a friend in cDc.

12   "In our circle": Interview with Brewer.

14   "Book of Cow": Franken Gibe, "The Book of Cow," 1987, http://textfiles
      .com/groups/CDC/book.of.cow.

14   "I took my stupidity very seriously": This is from a later text file, Fran-
      ken Gibe, "Retro Cow," 1989, www.cultdeadcow.com/cDc_files/cDc-0100
      .html.

14   "a decent summary of software commands": Franken Gibe, "Gibe's UNIX
      COMMAND Bible," 1987, http://textfiles.com/groups/CDC/cDc-0014.txt.

14   "telecom as a means, not an end": This phrase and close variations ap-
      peared in cDc files and public statements, including www.cultdeadcow
      .com/cDc_files/cDc-0100.html.

15   "No longer could this strong desire": Psychedelic Warlord, "Visions from
      the Last Crusade," 1988, www.textfiles.com/groups/CDC/visions/crusade.

15   "The first cDc file Warlord published": Psychedelic Warlord, "A Feature
      on MONEY—Today's Monster," 1987, http://textfiles.com/groups/CDC
      /cDc-0031.txt.

15   "interview with a self-proclaimed neo-Nazi": Psychedelic Warlord, "Inter-
      view with Neo-Nazi 'Ausderau,'" 1988, http://textfiles.com/groups/CDC
      /cDc-0059.txt.

18   "Chris Tucker, who dialed in from a board in Rhode Island": Chris Tuck-
      er's history comes from interviews with Osband, Mudge, Kevin, and others
      in cDc.

19   "In June 1971": The best account of the Yippie-phreaker coevolution is in
      Phil Lapsley's *Exploding the Phone* (New York: Grove Press, 2013).

20   "Political Rant #1": Nightstalker, "Political Rant #1," September 1, 1997,
      www.cultdeadcow.com/cDc_files/cDc-0339.txt.

Chapter 3: The Cons

21   "Houston-area hacker Jesse Dryden": I was unable to reach Jesse through close friends, relatives, database searches, or previous email addresses. This account of his career is built on interviews with his mother, former housemates, close friends, and members of cDc.

22   "better living through chemistry": The comment came in my interview with Mann. She also showed me an advance excerpt from her memoir, *The Band's with Me* (self-pub., Big Gorilla Books, 2018), epub.

24   "Jesse strategically leaked word": *Phrack* #32, November 17, 1990, www .phrack.org/issues/32/10.html.

25   "LoD began even before cDc, spawned in the early 1980s": For the history of the two groups and the trial of Neidorf, I am drawing on my own interviews with LoD and MoD members and others at the conferences. I also used Bruce Sterling's *The Hacker Crackdown* (New York: Bantam Books, 1992) and *Masters of Deception*, by Michelle Slatalla and Joshua Quittner (New York: HarperPerennial, 1995).

26   "Attendee Dale Drew of Arizona": Drew went on to have a serious security career with Tymnet, MCI, and Level 3 Communications, where he was chief security officer. He didn't respond to my interview request.

26   "Barlow's fellow acid-taking Deadhead": For more on Brand and the connections between psychedelics and major technology innovations, see John Markoff's *What the Dormouse Said* (New York: Viking, 2005).

27   "I've been in redneck bars wearing shoulder-length curls": John Perry Barlow, "Crime and Puzzlement," Electronic Frontier Foundation, June 1990, www.eff.org/pages/crime-and-puzzlement. The site has a collection of his other writings as well.

28   "Ladopoulos and Abene were arrested and prosecuted": One member of MoD who got away, Red Knight, was also in cDc. He later wrote to four cDc old-timers that after the arrests started, he quit hacking and went into the construction business.

28   "We were basically blacklisted": Goggans gave his account to *Gray Areas* magazine in 1994. He did not respond to my requests for comment, and neither did Chasin.

28   "At one HoHoCon, Goggans told an audience": His talk is recorded in a private film with highlights of the conference, which was shown to me by a cDc member.

30   "The reason I put on HoHoCon is": Jesse said this in a 1994 documentary by a woman using the name Annaliza Savage called Unauthorized Access, available here: https://archive.org/details/Hacker_Documentary _-_1994_-_Unauthorized_Access_by_Annaliza_Savage.

31   "He explained MindVox that year in an epic text file": Patrick Kroupa,
       "Voices in My Head," Excited Delirium, February 14, 1992, http://
       exciteddelirium.net/voices-in-my-head-mindvox-overture/.

32   "The general debauchery" was described by multiple eyewitnesses.

33   "already legendary to Moss": Both Moss and Bednarczyk told me this
       story.

34   "many of them were not true": As an example, his friend Angela Dor-
       mido told me that Jesse sent her a picture of Marilyn Manson's group and
       said he was on tour with Manson guitarist Jeordie White and the others.
       Dormido was friends with Waylon Jennings's son Shooter, a musician who
       eventually wound up on a tour bus with White. Shooter phoned Dormido
       and handed the phone to White: he had never heard of Jesse.

Chapter 4: Underground Boston

39   "One day in August": I interviewed a half-dozen attendees. Each detail I
       used was confirmed by at least two people. That was my general rule for
       this book, except for childhood memories and minor points.

40   "Brian and I had this vision": I am drawing on my own interview with John
       Lester and one he gave to *Decipher*, a blog hosted by Duo Security that ran
       a history of the L0pht in 2018. Dennis Fisher, "'We Got to Be Cool About
       This': An Oral History of the L0pht, Part I," *Decipher*, March 6, 2018,
       https://duo.com/decipher/an-oral-history-of-the-l0pht.

44   "Misha had followed the credo laid out by early hacker the Mentor": The
       Mentor, "The Conscience of a Hacker," *Phrack* #7, January 8, 1986,
       http://phrack.org/issues/7/3.html#article.

47   "participant Jordan Ritter": In addition to Ritter and Fanning, others in
       my Napster book *All the Rave* who show up in this volume are John Perry
       Barlow, Yobie Benjamin, Bill Gates, Steve Jobs, Jan Koum, Kevin Mit-
       nick, and Dug Song. Napster cofounder Sean Parker went on to serve
       as Facebook's first president, coaching Mark Zuckerberg through deal-
       ings with venture capitalists and helping him keep voting control of the
       company as it moved toward becoming one of the most important in the
       world.

50   "the *Boston Herald* identified New Hack City": Mark Mueller, "Hackers
       Go into Hiding as FBI Hunts for 'u4ea,'" *Boston Herald*, March 10, 1996.

Chapter 5: Back Orifice

54   "Mudge's list of aliases ran for ten pages": This is per Mudge, who does at
       times exaggerate.

54   "Byron York": York's history was described by Mudge and MacMillan and
       in some contemporaneous online reporting. His HoHoCon talk appears in

the private film of the event. I was unable to locate him. He is not the older man of the same name who has worked as a conservative writer for the *National Review*, *The Hill*, and other publications.

56   "There was one hitch": This section is based on interviews with multiple people who were there.

57   "Once, a leading security figure came to the L0pht": The figure was Marcus Ranum, who set up the first White House internet email and invented the modern intrusion-detection system. The anecdote and discussion of Mudge's dealings with malicious hacking come from my interviews with him in October 2018.

59   "Luke Benfey's 1994 *Dateline* interview": The interview has been transcribed by Misha, who changed Luke's name. That transcript is available here: www.cultdeadcow.com/oldskool/dateline.html.

60   "A 1996 story in the *San Antonio Express-News*": Chris Williams, "Air Force in Dogfight with Hackers," *San Antonio Express-News*, August 11, 1996. The same story ran in the *Rocky Mountain News* a week later under a different headline. Neither version is currently online.

60   "We intend to dominate and subvert the media": This statement appeared in cDc website updates including this one: www.cultdeadcow.com/news /medialist.htm.

60   "We're a neo-Marxist, anarcho-socialist guerrilla unit": Omega, "cDc Response to *Newsday* Magazine by Omega," December 1, 1996, https://w3 .cultdeadcow.com/cms/1996/12/cdcs-response-t.html.

60   "It's one thing if you have a state sponsor of disinformation": This came from hacker Mike Seery, who used the handle Reid Fleming. Seery was an old friend of Misha's and a longtime active cDc member credited by Misha for the neo-Marxist line.

61   "public spectacle to affect the public debate": The slogan is from a Yes Men page, http://yeslab.org/theyeslab.

64   "Would I be in trouble if I released a program that others could use to hack people?": The story comes from my interview with Josh.

67   "returned with an article on Back Orifice alone": Matt Richtel, "Hacker Group Says Program Can Exploit Microsoft Security Hole," *New York Times*, August 4, 1998, https://archive.nytimes.com/www.nytimes.com /library/tech/98/08/cyber/articles/04hacker.html.

67   "Microsoft is fully buzzword-compliant": The raw footage of this interview was provided to me by a cDc member.

68   "the local Atlanta field office of the FBI": Various memos and other FBI records were obtained through a Freedom of Information Act request by cDc members, who shared them with me but have not made them public.

69   "a comprehensive set of security features": Microsoft's original message is now gone from its site. cDc reposted it, with a point-by-point rebuttal, here: www.cultdeadcow.com/tools/bo_msrebuttal.html.

Chapter 6: One Million Dollars and a Monster Truck

71   "Kevin Wheeler sympathized": In an email to the group.

73   "*Wired* and the *Washington Post* had written about it": See, among other stories: Austin Bunn, "Beyond HOPE Hacks into Big Time," *Wired*, August 11, 1997, www.wired.com/1997/08/beyond-hope-hacks-into-big-time, and Pamela Ferdin, "Into the Breach," *Washington Post*, April 4, 1998, www.washingtonpost.com/archive/politics/1998/04/04/into-the-breach/8ae3cf86-fbd7-4037-a1b6-842df39d9db7.

74   "The success of Eligible Receiver": For more on Eligible Receiver and Moonlight Maze, see Fred Kaplan, *Dark Territory* (New York: Simon & Schuster, 2016), and Thomas Rid, *Rise of the Machines* (New York: W. W. Norton, 2016).

75   "Clarke took a crew from the NSC": Different members of the L0pht tell slightly different versions of how Clarke came to hear about and visit the L0pht and how the testimony was arranged. I am going with what Clarke told me about finding them.

75   "If you have an offer, we'll listen": The joke is by Mudge's recollection. The others recall the part about Clarke being surprised the L0pht could do what it did without a government's support.

76   "Mudge told the senators": Cris "Space Rogue" Thomas, the best archivist of the L0pht's members, posted a transcript of the hearing here: www.spacerogue.net/wordpress/?p=602.

76   "a problem they had found in the internet's routing procedure, Border Gateway Protocol": It has never been made clear what bug the group was referring to. Mudge said at a L0pht reunion panel at Def Con 2018 that he had found it on his day job at BBN. He told me it had just been reported to router makers before the testimony.

76   "We were a visceral representation of what the adversarial view was": Wysopal's comment came during the 2018 Def Con panel marking the twentieth anniversary of the testimony.

78   "The Atlanta FBI office warned the Pentagon": The FBI records were obtained through a Freedom of Information Act request by cDc members, who have not made them public but shared them with me.

78   "The Defense Department's Criminal Investigative Service": According to declassified CIS documents shown to me.

79   "A lawyer was hired": cDc member Mike Seery put up the $1,000 needed. The lawyer was Cindy Cohn.

79   "An ISS intermediary even offered cash": According to a log of the Internet Relay Chat, which is not publicly available. The man said in the chat that he worked for ISS at the time, though his LinkedIn profile shows he joined full-time in 2000.

79   "ISS is just flat-out sleazy in a lot of ways": Mudge said this to a filmmaker at the time. I have seen the footage.

80   "one million dollars and a monster truck": The letter, signed with Mike Seery's handle, was cited by the BBC and others. The full text is at www .mail-archive.com/siglinux@locutus.csres.utexas.edu/msg04587.html.

81   "Christien had burned advance copies of BO2k": The story of how the CDs came to be infected was told to me by Christien and other cDc members. Fried declined interview requests.

82   "practically calling us godless commies": Kevin's comment was in an email to others in cDc. The paper's editorial ran on July 15, 1999. It is not currently online.

82   "a qualified thumbs-up": Bruce Schneier, "Back Orifice 2000," *Crypto-Gram* (newsletter), *Schneier on Security* (blog), August 15, 1999, www .schneier.com/crypto-gram/archives/1999/0815.html#BackOrifice 2000.

83   "One Lockheed Martin expert wrote to a security mailing list": The email went to subscribers of the list called NTBugtraq.

83   "Carrie wanted to help Microsoft do better": My sources for this anecdote are Carrie and Beck.

84   "the leading tech discussion site Slashdot": "Bizarre Answers from Cult of the Dead Cow," Slashdot, October 22, 1999, https://news.slashdot.org /story/99/10/22/1157259/bizzare-answers-from-cult-of-the-dead-cow.

Chapter 7: Oxblood

85   "John Lester's personal account": Count Zero, "HoHoCon 1994 . . . The Insanity Continues," January 6, 1995, www.cultdeadcow.com/oldskool /HoHo94.html.

86   "Laird said he was working for a not-for-profit": He later told me he had been volunteering at the Toronto group Web Networks, which built websites for progressive groups, native tribes, and government agencies, and supported himself with other jobs on the side.

86   "Laird came by his sense of ethics": I feel obliged to remind readers that, as with Mudge and the others, I am relying on Laird's own word for this account of his pre-cDc life.

88   "Laird memorialized the event in classic cDc style": This was in an email circulated to the group.

89   "A Declaration of the Independence of Cyberspace": John Perry Barlow, "A Declaration of the Independence of Cyberspace," Electronic Frontier Foundation, February 8, 1996, www.eff.org/cyberspace-independence.

90   "Barlow said that the innocence": I interviewed him in a San Francisco nursing home near the end of his life.

93   "a short piece in *Wired* magazine about the Blondes": Arik Hesseldahl,
        "Hacking the Great Firewall," *Wired*, December 1997, 120, www.scribd
        .com/doc/237686960/Hacking-the-Great-Firewall.

94   "Laird wrote that the conversation had taken place": Oxblood Ruffin,
        "The Longer March," July 15, 1998, www.cultdeadcow.com/cDc_files/cDc
        -0356.html.

95   "As leader of the Hong Kong Blondes": Arik Hesseldahl, "Hacking for
        Human Rights?," *Wired News*, July 14, 1998, www.cultdeadcow.com/news
        /wired/19980714/.

95   "Clinton had been working to normalize relations": "President Clinton's
        Visit to China in Context," Human Rights Watch, n.d., www.hrw.org/legacy
        /campaigns/china-98/visit.htm.

95   "Klein's wide-eyed write-up": Naomi Klein, "Computer Hacking New
        Tool of Political Activism," *Toronto Star*, July 23, 1998, reprinted at www
        .cultdeadcow.com/news/newspapers/toronto_star72398.txt. Klein also
        wrote about the Blondes in her book *No Logo*, in which she explained that
        she had confirmed the legitimacy of the Laird-Wong interview with the
        "subject" of that piece. Klein declined repeated interview requests.

96   "Was releasing Back Orifice to the public immoral?": "St. Paul, Back
        Door Boom Boom, and All the Tea in China" (press release), August 6,
        1998, http://cultdeadcow.com/news/response.txt.

97   "a respected China-based writer for the *Los Angeles Times* wrote a front-
        page feature story": Maggie Farley, "Dissidents Hack Holes in China's
        New Wall," *Los Angeles Times*, January 4, 1999, http://articles.latimes
        .com/1999/jan/04/news/mn-60340.

98   "he said he had met Wong at a party": Oxblood Ruffin, "Chinese Check-
        ers," cDc text file #361, December 23, 1998, www.cultdeadcow.com/cDc
        _files/cDc-0361.html.

98   "cDc issued a joint statement": "LoU Strike Out with International Coa-
        lition of Hackers: A Joint Statement by *2600*, the Chaos Computer Club,
        the Cult of the Dead Cow, !Hispahack, L0pht Heavy Industries, *Phrack*
        and Pulhas" (press release), January 7, 1999, www.cultdeadcow.com
        /news/statement19990107.html.

98   "The LoU, which had been split internally over the matter": A member
        of LoU told Misha the internal story during a panel for a screening of the
        documentary on Anonymous, *We Are Legion*. LoU member Bronc Buster
        later joined Hacktivismo and worked on an early, rough version of Peeka-
        booty, a privacy-protecting browser.

99   "Laird walked the tale halfway back": Oxblood Ruffin, "Blondie
        Wong and the Hong Kong Blondes," Medium, March 23, 2015, https://
        medium.com/emerging-networks/blondie-wong-and-the-hong-kong
        -blondes-9886609dd34b.

101   "Hacktivismo Declaration": The entire declaration was disseminated within a joint cDc-Hacktivismo press release: "International Bookburning in Progress," July 4, 2001, www.cultdeadcow.com/cDc_files/declaration .html.

102   "I didn't write the 'Harlem Declaration' to preach to the converted": This was in an email Laird sent others in cDc.

102   "In a public FAQ post": "The Hacktivismo FAQ v1.0," 2000–2001, www .cultdeadcow.com/cDc_files/HacktivismoFAQ.html.

102   "Milošević, acting as his own attorney": Ball's cross-examination is available on the website of the International Criminal Tribunal for the Former Yugoslavia. The Cult of the Dead Cow question came on March 14, 2002, at page 2228 of the trial transcript. www.icty.org/x/cases/slobodan _milosevic/trans/en/020314IT.htm.

104   "The program, informally known as 'internet in a box'": Alexander Howard, "Exit Interview: Alec Ross on Internet Freedom, Innovation and Digital Diplomacy," Huffington Post, March 12, 2013, www.huffingtonpost .com/alexander-howard/exit-interview-alec-ross-_b_2860211.html.

104   "Adam O'Donnell, known as Javaman, also worked on a CIA project": The section on O'Donnell's CIA work is based on interviews with two people familiar with it.

Chapter 8: Much @stake

110   "overexcited public relations people told media the real names": Wysopal recalls that the first outlet to publish their names was *Newsweek*. Mudge says he was also outed by the White House, which put him on a list of those meeting the president.

110   "having sex with a prostitute": Three senior @stake employees independently told me the story.

111   "She lost the vote and a few days later was proven right": Snyder is my main source for the account of her Microsoft tenure.

111   "testing the security of an SQL database for a German bank": Litchfield told the story himself in an article on Threatpost: David Litchfield, "The Inside Story of SQL Slammer," Threatpost, October 20, 2010, https:// threatpost.com/inside-story-sql-slammer-102010/74589/.

112   "a 2003 paper arguing that Microsoft's monopoly was bad for security": Dan Geer et al., "Cyber*In*security: The Cost of Monopoly," http://geer .tinho.net/pubs.

114   "an intelligence contractor I will call Rodriguez": This story comes from multiple interviews with Rodriguez.

114   "location tracking in every cell phone": The defense of location privacy, Hong Kong Blondes admission, and lone-wolf stories come from Mudge.

116   "Ultimately, I just cracked a bit": Mudge's first public admission of his
          mental health issues came in a good 2015 *Washington Post* series about
          why the internet's security flaws remain unfixed: Craig Timberg, "A Di-
          saster Foretold—and Ignored," *Washington Post*, June 22, 2015, www
          .washingtonpost.com/sf/business/2015/06/22/net-of-insecurity-part-3/.

117   "Ninja Strike Force member I will call Stevens": Stevens told both me and
          another source of his experiences.

117   "Some operatives installed keyloggers": This was reported in Sean Nay-
          lor's recent history of JSOC, *Relentless Strike* (New York: St. Martin's
          Press, 2015).

117   "Others had similar experiences": Thieme provided me with the emails
          from veterans.

120   "The first mainstream articles on the zero-day business": Andy Greenberg
          profiled the @stake veteran who calls himself the Grugq in "Shopping for
          Zero-Days: A Price List for Hackers' Secret Software Exploits," *Forbes*,
          March 23, 2012, www.forbes.com/sites/andygreenberg/2012/03/23/
          shopping-for-zero-days-an-price-list-for-hackers-secret-software
          -exploits/. I later wrote a deeper story and a sidebar for Reuters: "Special
          Report: U.S. Cyberwar Strategy Stokes Fear of Blowback," Reuters, May
          10, 2013, www.reuters.com/article/us-usa-cyberweapons-specialreport
          /special-report-u-s-cyberwar-strategy-stokes-fear-of-blowback-idUSBRE
          9490EL20130510, and "Booming 'Zero-Day' Trade Has Washington Cy-
          ber Experts Worried," Reuters, May 10, 2013, www.reuters.com/article
          /us-usa-cyberweapons-policy/booming-zero-day-trade-has-washington
          -cyber-experts-worried-idUSBRE9490EQ20130510.

120   "organized criminals, a preponderance of them in Russia and Ukraine":
          I cover the evolution of botnets and the reason for Russian prominence in
          malware in *Fatal System Error* (New York: PublicAffairs, 2010).

121   "once you accept that there are bugs": "Canvassing All Security Cracks,"
          Sydney Morning Herald, April 22, 2005, www.smh.com.au/technology
          /canvassing-all-security-cracks-20050422-gdl620.html. Aitel did not re-
          spond to my interview requests.

121   "They rejected illegal jobs": Interview with Val Smith.

122   "Project Mayhem": "Phrack Prophile on the UNIX Terrorist," *Phrack* #65,
          November 4, 2008, http://phrack.org/issues/65/2.html.

124   "the new consulting firm, iSec Partners": The story of iSec comes from my
          interviews with Stamos and an electronic exchange with Rubin.

Chapter 9: Tor and Citizen Lab

128   "Hacktivismo is good with thinking up new projects": Robert Lemos, "Long
          Haul Ahead for Social Hackers," ZDNet, February 19, 2002, www.zdnet

.com/article/long-haul-ahead-for-social-hackers/. Baranowski declined my interview requests. DeVilla spoke in an interview with me.

132   "Some of our early interactions around hacktivism": Deibert gives Laird Brown credit not only in his comments to me but also in his book *Black Code* (Toronto: Signal, 2013).

133   "in the context of international security": The early scope is described in Deibert's book *Black Code*.

133   "Silicon Valley firm Blue Coat": The Blue Coat research drew main-stream-media attention. The company blamed resellers of its products.

133   "The lab also took on the legal sale of exploits": The lab's research is highlighted on its website: https://citizenlab.ca/category/research/.

133   "A devastating series of four front-page reports in the *New York Times*": For example, see Azam Ahmed, "Spyware Trailed Investigators in Mex-ico," *New York Times*, July 9, 2017, www.nytimes.com/2017/07/10/world /americas/mexico-missing-students-pegasus-spyware.html.

134   "But they faced accusations of bias": I wrote about the Balkanization of high-end security research in "Politics Intrude as Cybersecurity Firms Hunt Foreign Spies," Reuters, March 11, 2015, www.reuters.com/article/us -cybersecurity-fragmentation-insight/politics-intrude-as-cybersecurity -firms-hunt-foreign-spies-idUSKBN0M809N20150312.

135   "Deibert's team dubbed the spy network GhostNet": The original Ghost-Net report—"Tracking GhostNet: Investigating a Cyber Espionage Net-work," March 28, 2009—is here: https://issuu.com/citizenlab/docs/iwm -ghostnet.

Chapter 10: Jake

140   "He also had an extraordinarily compelling personal story": A number of journalists have recounted Appelbaum's upbringing, including Nathan-iel Rich in a *Rolling Stone* article ("The American Wikileaks Hacker," December 1, 2010, www.rollingstone.com/culture/culture-news/the -american-wikileaks-hacker-238019/). One longtime friend of Jake's vouched for the major points in the *Rolling Stone* story. Appelbaum him-self did not respond to my interview requests by email, Twitter direct mes-sage, and emails to his graduate school advisors.

142   "a bizarro version of Mark Zuckerberg": Rich, "The American Wikileaks Hacker."

142   "Even more of a show-off than Jake": The best work on Assange is Andy Greenberg's book *This Machine Kills Secrets* (New York: Plume, 2012). His emails to the Cypherpunks list are available on the list archive, which tends to move around a bit online.

145   "The story of Anonymous": See Gabriella Coleman, *Hacker, Hoaxer,*

Whistleblower, Spy (Brooklyn, NY: Verso, 2014); and Parmy Olson, *We Are Anonymous* (New York: Back Bay Books, 2012).

147   "I wrote a short 2011 story in the *Financial Times*": "Cyberactivists Warned of Arrest," *Financial Times*, February 4, 2011, www.ft.com/content /87dc140e-3099-11e0-9de3-00144feabdc0. My other stories on Anonymous and LulzSec included "They're Watching, and They Can Bring You Down," *FT Magazine*, September 23, 2011, www.ft.com/content/3645ac3c -e32b-11e0-bb55-00144feabdc0#axzz1YtFTuZd2.

147   "What we did was different": Ryan Gallagher, "Why Hacker Group LulzSec Went on the Attack," *Guardian*, July 14, 2011, www.theguardian .com/technology/2011/jul/14/why-lulzsec-decided-to-disband.

147   "Davis later said": In an email conversation with me.

148   "Assange was tracking events closely": Olson, *We Are Anonymous*, 326–329.

148   "Russia also had a substantial presence": UK and US law enforcement officials told me this as I was covering Anonymous for the *Financial Times*. I have interviewed Cassandra Fairbanks and noted her curious evolution for Reuters.

150   "WikiLeaks's flagging reputation": How Snowden chose his journalists was laid out long after he went public. This version was presented at a memorial for John Perry Barlow, which I attended. A video of the memorial is available online and worth watching: https://supporters.eff.org/civicrm /event/info?reset=1&id=191. Trevor Timm talked about the release of the Snowden documents; the discussion begins at around 1:32:00 of the video.

151   "Jake later reported related stories for *Der Spiegel*": The heart of these stories is what is known as the ANT catalog, which details specific attacks. The *Guardian* and other publications generally shied away from identifying the devices and software the NSA could hack.

151   "Other stories showed that the NSA had continued to corrupt security products": Good accounts of the NSA subverting standards, under a project called Bullrun, include these: Nicole Perlroth, Jeff Larson, and Scott Shane, "N.S.A. Able to Foil Basic Safeguards of Privacy on Web," *New York Times*, September 5, 2013, www.nytimes.com/2013/09/06/us /nsa-foils-much-internet-encryption.html; James Ball, Julian Borger, and Glenn Greenwald, "Revealed: How US and UK Spy Agencies Defeat Internet Privacy and Security," *Guardian*, www.theguardian.com/world/2013 /sep/05/nsa-gchq-encryption-codes-security; and "Dual EC DRBG," Project Bullrun, July 31, 2005, https://projectbullrun.org/dual-ec/index.html.

152   "Song urged Koum": These details come from three people with knowledge of the events.

152   "citing the opportunity": Acton's initial statement is here: https://signal .org/blog/signal-foundation/. The second quote is from an interview with me.

153   "He bragged of multiple lovers": Poitras acknowledged the relationship in her film *Risk*. Jardin confirmed her relationship by email.

153   "Steele came too late for some": Komlo wrote her account anonymously for the protest website JacobAppelbaum.net, then later came forward by name. I spoke to her after that. I also interviewed Leigh Honeywell and others involved in the Tor investigation. As stated earlier, Appelbaum did not respond to interview requests. Neither did Bernstein. Gilmore's early defense came on a private email list.

154   "Being involved with him was a steady stream of humiliations small and large": Leigh Honeywell, "He Said, They Said" (blog post), hypatia.ca, June 7, 2016, https://hypatia.ca/2016/06/07/he-said-they-said/.

155   "What you tolerate and don't tolerate defines you": This is from an interview with someone involved in the investigation.

157   "You can't dialogue with a sociopath": Farr wrote this as a post on Medium. He later deleted it, saying that he did not want to further divide the security community. An archive of the original is here: https://web .archive.org/web/20160606222408/https://medium.com/@nickf4rr /hi-im-nick-farr-nickf4rr-35c32f13da4d.

157   "most serious public statement in more than a decade": "CULT OF THE DEAD COW Statement on Jacob Appelbaum / ioerror" (press release), June 6, 2016, http://w3.cultdeadcow.com/cms/2016/06/cult-of-the-dead -cow-statement-on-jacob-appelbaum-ioerror.html.

158   "In a personal post on Medium": Oxblood Ruffin, "Public Figures & Anonymous Victims," Medium, June 8, 2016, https://medium.com /@oxbloodruffin/public-figures-anonymous-victims-543f0b02d684.

159   "quote the emails between WikiLeaks and its real source": "Read Mueller Probe Indictment of 12 Russians for Hacking Democrats," *Washington Post*, n.d., http://apps.washingtonpost.com/g/documents/national/read -mueller-probe-indictment-of-12-russians-for-hacking-democrats /3087/.

Chapter 11: Mixter, Muench, and Phineas

162   "When I was young, there was something fun": Marlinspike's comments came in a really good *Wired* profile by Andy Greenberg: "Meet Moxie Marlinspike, the Anarchist Bringing Encryption to All of Us," *Wired*, July 31, 2016, www.wired.com/2016/07/meet-moxie-marlinspike -anarchist-bringing-encryption-us/.

162   "an early supporter of Laird's Hacktivismo project named Martin Muench": Muench did not respond to my interview requests.

163   "sixty-page Gamma catalog": A partial version is online at https://archive .org/stream/186_201106-ISS-ELAMAN1/186_201106-ISS-ELAMAN1 _djvu.txt.

165   "I just read the Citizen Lab reports": Lorenzo Franceschi-Bicchierai, "Hacker 'Phineas Fisher' Speaks on Camera for the First Time—Through a Puppet," Motherboard, July 20, 2016, https://motherboard.vice.com /en_us/article/78kwke/hacker-phineas-fisher-hacking-team-puppet. The interview was conducted by VICE reporter Lorenzo Franceschi-Bicchierai, who did the best work on Gamma's hacking and several copycat attacks on spyware vendors. Not unreasonably, he declined to pass along my interview request to Phineas, whom I was unable to reach.

165   "In another interview a month earlier": Enric Borràs, "Phineas Fisher; 'I'm Wanted by Much More Powerful Police Forces than Catalonia's and for Much Worse Crimes," Ara, June 6, 2016, www.ara.cat/en/Im-much -powerful-Catalonias-crimes_0_1590441016.html. The author of that article also declined to pass along my interview request to Phineas.

166   "If you're a spouseware vendor, we're coming for you": The group posted its widely quoted warning and advice on Pastebin: https://pastebin.com /raw/Y1yf8kq0.

166   "public interest hacking": Gabriella Coleman, "The Public Interest Hack," Limn, issue 8 (February 2017), https://limn.it/articles/the -public-interest-hack/.

166   "articles about the leaked Panama Papers": The work was led by the International Consortium of Investigative Journalists (www.icij.org), with the McClatchy newspaper chain and the Miami Herald playing major roles.

168   "Two former Kaspersky employees told me": When I asked Eugene Kaspersky about the claims, he acknowledged his software sometimes took inactive code. Joseph Menn, "Kaspersky Acknowledges Taking Inactive Files in Pursuit of Hackers," Reuters, November 3, 2017, www .reuters.com/article/us-cyber-summit-kaspersky/kaspersky-acknowledges -taking-inactive-files-in-pursuit-of-hackers-idUSKBN1D328B.

169   "I once again express my sympathy and profound condolences to the family of the Russian pilot": Alec Luhn and Ian Black, "Erdoğan Has Apologised for Downing of Russian Jet, Kremlin Says," Guardian, June 27, 2016, www.theguardian.com/world/2016/jun/27 /kremlin-says-erdogan-apologises-russian-jet-turkish.

169   "I hacked AKP": Dissent, "Notorious Hacker 'Phineas Fisher' Says He Hacked Turkey's Ruling Political Party," July 21, 2016, https://www .databreaches.net/notorious-hacker-phineas-fisher-says-he-hacked -turkeys-ruling-political-party/.

170   "UK security activist Thomas White": White later removed his personal site from the web.

170   "Phineas told VICE he was retiring": Lorenzo Franceschi-Bicchierai, "Hacking Team Hacker Phineas Fisher Is Taking a Break Because of

Stress," *Motherboard*, February 9, 2017, https://motherboard.vice.com /en_us/article/xy5enw/hacking-teams-phineas-fisher-will-return-but-only -after-a-break-at-the-beach.

172   "collaborative reasoning tool": Pitsos described Kialo that way to the *Financial Times* in "Meet the Start-Up That Wants to Sell You Civilised Debate," January 24, 2018, www.ft.com/content/4c19005c-ff5f -11e7-9e12-af73e8db3c71.

Chapter 12: Mudge and Dildog

175   "Peiter Zatko, known to even close friends as Mudge": There are multiple stories about how Mudge took his best-known handle. The truth is the most boring one: It was a classmate's actual last name, as Mudge explained to tech journalist Elinor Mills in a taped interview.

175   "it got loose in its test version": Interview with Cerf.

177   "the creation and suppression of strategic surprise": Dugan used this version of the phrase in various talks, but it dates to at or near the agency's creation. Similar wording is in a DARPA fact sheet here: www.darpa.mil /attachments/DARPA_Fact_Sheet_1_07-25-17.pdf.

177   "Now he called in a dozen": My main sources for the meeting are Song and Mudge. Mudge also credited Song with the CFT idea in a talk on YouTube.

178   "Miller was presenting": The story of Miller's funding comes from both Miller and Mudge.

179   "Cyber Analytic Framework": Parts of the Framework are classified, but Mudge has discussed other aspects of it with me and in talks available on YouTube. It has been reported elsewhere that another project of Mudge's, to detect unusual activity on a network, was aimed at ferreting out moles and whistle-blowers. But Mudge vigorously disputes that, saying that it hunted for actions by user credentials being wielded by outsiders. Kaufman backs Mudge's version.

181   "Mudge accepted the secretary of defense's highest award for civilian service": I saw a hand-redacted version of the citation.

182   "a secure operating system on a memory card": Mudge talked about the project at Google's annual developer's conference in 2015; the talk can be viewed here: www.youtube.com/watch?v=mpbWQbkl8_g.

182   "a harder time attacking Google's Chrome browser": Mudge and Sarah Zatko have released various findings from the lab in talks at Black Hat and other conferences.

185   "I hate Adobe": A large proportion of criminal and geopolitical malware depended on Flash vulnerabilities for years. The bad security was one of the reasons that Steve Jobs killed Apple support for it. In 2018, Flash is nearing end of life.

187   "Gallagher gave him a shout-out": Hugh Gallagher, "White Boy Rocks Harlem," posted by zpin, YouTube video, 2:40, June 28, 2006, www.youtube.com/watch?v=Hv1ihFI5iKI.

188   "In four years, the group found 1,400 vulnerabilities": Figures disclosed by Project Zero and Google Chrome overseer Parisa Tabriz at her Black Hat keynote in 2018, covered here: Seth Rosenblatt, "Google's 'Security Princess' Calls for Stronger Collaboration," Parallax, August 8, 2018, www.the-parallax.com/2018/08/08/google-security-princess-parisa-tabriz-black-hat/.

Chapter 13: The Congressman and the Trolls

189   "a punk band, Foss": The band also featured Cedric Bixler-Zavala, later lead singer of Grammy Award–winning the Mars Volta. Here's Foss on a television show in El Paso in 1994: "Foss on Let's Get Real TV show- El Paso, TX- 1994 Pt 3- The Song," posted by elephantandseal, YouTube video, 9:59, June 30, 2012, www.youtube.com/watch?time_continue=2&v=eI5GGPFnX24.

189   "one of the poorest cities in America": And still eighth-poorest several years later, per a CBS News ranking in February 2015: Bruce Kennedy, "America's 11 Poorest Cities," MoneyWatch, CBS News, February 18, 2015, www.cbsnews.com/media/americas-11-poorest-cities/.

189   "a slim book": Beto O'Rourke and Susie Byrd, *Dealing Death and Drugs: The Big Business of Dope in the U.S. and Mexico* (El Paso, TX: Cinco Puntos Press, 2011).

190   "He showed voters the energy he could devote": There are many decent accounts of Beto's career and campaign, though none picked up on his early hacking and bulletin-board posts. Among the better stories are Patrick Svitek, "Rep. Beto O'Rourke, in Long-Shot Bid for Senate, Is No Stranger to 'Calculated Risks,'" *Texas Tribune*, April 7, 2017, www.texastribune.org/2017/04/07/beto-orourke-2018-senate-bid-ted-cruz/; and Eric Benson, "What Makes Beto Run?," *Texas Monthly*, January 2018, www.texasmonthly.com/articles/makes-beto-orourke-run/.

190   "Beto began broadcasting the event from his phone over Facebook": Allana Akhtar and Paul Singer, "Facebook Live, Periscope Have Big U.S. Political Moment with House Sit-In," *USA Today*, June 23, 2016, www.usatoday.com/story/tech/news/2016/06/23/facebook-live-periscope-have-big-political-moment-house-sit-/86297956/.

190   "they streamed live video.": Large segments of the livestream are findable with the hashtag #BipartisanRoadtrip.

193   "Never Again" pledge: https://neveragain.tech.

195   "Speaking as Matlock": After my interview with him, an antifascist group published Matlock's real name. Two of Noonan's associates then confirmed

it to me. In 2019, Noonan told me he had moved on: "I'm out of politics and I have been getting far-right activists and white nationalists, many of whom I was with at Charlottesville, to drop acid and slam ketamine in an effort to reevaluate their lives and stay relevant to society instead of going down the autistic rabbit hole."

195   "But Auernheimer was suspected of hosting faked documents": Eric Geller, "Neo-Nazi Activist May Be Behind Fake Macron Accounts," Politico, January 28, 2018, www.politico.eu/article/neo-nazi -activist-may-be-behind-fake-macron-documents/. In a 2019 email exchange with me, Auernheimer declined to answer questions about his activities in the French or American elections but said he did not work with Russia. He did work at times with right-wing troll Charles "Chuck" Johnson, whose startup WeSearchr coordinated bounty offers for the fruits of political opposition research, including "proof" Macron was gay and Clinton's deleted emails.

196   "network of companies including Cambridge Analytica": Coverage of Cambridge Analytica, including the identification of a whistle-blower, was led by the *Guardian*.

197   "I revealed that security company RSA had taken $10 million": "Exclusive: Secret Contract Tied NSA and Security Industry Pioneer," Reuters, December 20, 2013, www.reuters.com/article/us-usa-security-rsa-idUS-BRE9BJ1C220131220. A follow-up is here: Joseph Menn, "Exclusive: NSA Infiltrated RSA Security More Deeply than Thought—Study," Reuters, March 31, 2014, www.reuters.com/article/us-usa-security-nsa-rsa /exclusive-nsa-infiltrated-rsa-security-more-deeply-than-thought -study-idUSBREA2U0TY20140331?irpc=932.

198   "Stamos quit Yahoo in 2015": Joseph Menn, "Exclusive: Yahoo Secretly Scanned Customer Emails for U.S. Intelligence—Sources," Reuters, October 4, 2016, www.reuters.com/article/us-yahoo-nsa-exclusive /exclusive-yahoo-secretly-scanned-customer-emails-for-u-s-intelligence -sources-idUSKCN1241YT.

199   "Facebook experts found GRU reconnaissance of campaign workers": Joseph Menn, "Exclusive: Russia Used Facebook to Try to Spy on Macron Campaign," Reuters, July 26, 2017, www.reuters.com/article/us-cyber -france-facebook-spies-exclusive/exclusive-russia-used-facebook-to-try -to-spy-on-macron-campaign-sources-idUSKBN1AC0EI. I covered Facebook, propaganda, and hacking closely during this time and routinely interviewed intelligence, congressional, Facebook, and outside security sources.

199   "intelligence officials told *Time* magazine that Russian propagandists bought Facebook ads": Massimo Calabresi, "Inside Russia's Social Media War on America," *Time*, May 18, 2017, http://time.com/4783932 /inside-russia-social-media-war-america/.

199 "2018 indictments of thirteen Russians": Matt Apuzzo and Sharon
     LaFraniere, "13 Russians Indicted as Mueller Reveals Effort to Aid
     Trump Campaign," *New York Times*, February 16, 2018, https://www.ny
     times.com/2018/02/16/us/politics/russians-indicted-mueller-election
     -interference.html.

200 "Cruz also got tweets of support from IRA accounts": Josh Russell, "If
     you go look at the Clemson researchers database there are at least 4500
     tweets containing 'Cruz' dating all the way back to february 2015," Twit-
     ter, September 13, 2018, 7:39 p.m., https://twitter.com/josh_emerson
     /status/1040429696792637440.

200 "Stamos was trying to do the right thing": The board incident was reported
     in Sheera Frenkel et al., "Delay, Deny and Deflect: How Facebook's
     Leaders Fought Through Crisis," *New York Times*, November 14, 2018,
     www.nytimes.com/2018/11/14/technology/facebook-data-russia-election
     -racism.html.

201 "farewell memo": Ryan Mac and Charlie Warzel, "Departing Facebook
     Security Officer's Memo: 'We Need to Be Willing to Pick Sides,'" Buzz-
     Feed News, July 24, 2018, www.buzzfeednews.com/article/ryanmac
     /facebook-alex-stamos-memo-cambridge-analytica-pick-sides.

201 "Rank-and-file tech workers": Daisuke Wakabayashi and Scott Shane,
     "Google Will Not Renew Pentagon Contract That Upset Employees," *New
     York Times*, June 1, 2008, www.nytimes.com/2018/06/01/technology
     /google-pentagon-project-maven.html.

202 "Cruz is a rare and precious gift": Frank Bruni, "Watch Out, Ted Cruz,
     Beto Is Coming," *New York Times*, April 7, 2018, www.nytimes.com
     /2018/04/07/opinion/sunday/ted-cruz-beto-orourke-texas.html.

202 "*Vanity Fair* had dubbed him Kennedyesque": Abigail Tracy, "Meet
     the Kennedyesque Democrat Trying to Beat Ted Cruz," *Vanity Fair*,
     May 31, 2017, https://www.vanityfair.com/news/2017/05/beto-orourke
     -ted-cruz-texas-senate-2018.

202 "Beto told National Public Radio": Wade Goodwyn, "Texas Democrat's
     Underdog Bid to Unseat Ted Cruz Picks Up Momentum," *All Things
     Considered*, NPR, March 5, 2018, www.npr.org/2018/03/05/590709857
     /texas-democrats-underdog-bid-to-unseat-ted-cruz-picks-up-momentum.

203 "sons of bitches": Adam Edelman, "Trump Rips NFL Players After
     Anthem Protests During Preseason Games," NBC News, August 10,
     2018, www.nbcnews.com/politics/donald-trump/trump-rips-nfl-players
     -after-protests-during-preseason-games-n899551.

203 "Beto, who had never been asked the question before": Daniel Kreps,
     "Watch Beto O'Rourke Talk Trump's Texas Visit, NFL Kneeling Viral
     Video on 'Ellen,'" *Rolling Stone*, September 5, 2018, www.rollingstone

.com/politics/politics-news/watch-beto-orourke-talk-trumps-texas-visit-nfl-kneeling-viral-video-on-ellen-719245/.

204   "O'Rourke offers not just a path to victory in Texas": Peter Hamby, "'It Seems Like Iowa in 2007': Is Beto O'Rourke the Left's Obama-Like Answer to Trump in 2020?," *Vanity Fair*, August 29, 2018, www.vanityfair.com/news/2018/08/could-beto-orourke-be-the-next-obama.

204   "when I told him I wanted to include his background in a post-election book": Knowing a Congressman had belonged to the group, I guessed it was Beto from press coverage of his Senate race that described his rebellious youth in Texas. But other members would not confirm my suspicion, so I offered my word that I would not publish until after the November 2018 election. They agreed to my terms, and I then offered the same deal to Beto.

206   "the beginning of the end for one-party rule": James Henson, "Beto O'Rourke Should Run for Senate in 2020. He Could Win," *Washington Post*, November 9, 2018, https://www.washingtonpost.com/opinions/beto-orourke-should-run-for-senate-in-2020-he-could-win/2018/11/09/99263192-e462-11e8-ab2c-b31dcd53ca6b_story.html?utm_term=.d75abaa157b8.

207   "at the center of the dispute is Rep. Beto O'Rourke": Jonathan Martin and Alexander Burns, "Democrats Have Two Paths for 2020: Daring or Defensive. Can They Settle on Either?," *New York Times*, November 10, 2018, https://www.nytimes.com/2018/11/10/us/politics/democrats-2020-president.html.

207   "A month later": Matt Flegenheimer and Jonathan Martin, "Beto O'Rourke Emerges as the Wild Card of the 2020 Campaign-in-Waiting," *New York Times*, December 9, 2018, www.nytimes.com/2018/12/09/us/politics/beto-2020-presidential-race.html.

Epilogue

210   "Institute of Electrical and Electronic Engineers": The IEEE code is available at www.ieee.org/about/corporate/governance/p7-8.html.

211   "Security is about how you configure power": Song's speech was on YouTube for a time.

> INDEX

An investigative reporter for Reuters, **Joseph Menn** is the longest-serving and most respected mainstream journalist on cybersecurity. He has won three Best in Business awards from the Society of American Business Editors and Writers and been a finalist for three Gerald Loeb Awards. His *Fatal System Error: The Hunt for the New Crime Lords Who Are Bringing Down the Internet* exposed the Russian government's collaboration with organized criminal hackers and was named one of the ten best nonfiction books of 2010 by Hudson Booksellers. He also wrote the definitive *All the Rave: The Rise and Fall of Shawn Fanning's Napster*, an Investigative Reporters and Editors finalist for book of the year. He previously worked for the *Financial Times*, *Los Angeles Times*, and Bloomberg, and he has spoken at conferences including Def Con, Black Hat, and RSA. He grew up near Boston and lives in San Francisco.

PublicAffairs is a publishing house founded in 1997. It is a tribute to the standards, values, and flair of three persons who have served as mentors to countless reporters, writers, editors, and book people of all kinds, including me.

I. F. STONE, proprietor of *I. F. Stone's Weekly*, combined a commitment to the First Amendment with entrepreneurial zeal and reporting skill and became one of the great independent journalists in American history. At the age of eighty, Izzy published *The Trial of Socrates*, which was a national bestseller. He wrote the book after he taught himself ancient Greek.

BENJAMIN C. BRADLEE was for nearly thirty years the charismatic editorial leader of *The Washington Post*. It was Ben who gave the *Post* the range and courage to pursue such historic issues as Watergate. He supported his reporters with a tenacity that made them fearless and it is no accident that so many became authors of influential, best-selling books.

ROBERT L. BERNSTEIN, the chief executive of Random House for more than a quarter century, guided one of the nation's premier publishing houses. Bob was personally responsible for many books of political dissent and argument that challenged tyranny around the globe. He is also the founder and longtime chair of Human Rights Watch, one of the most respected human rights organizations in the world.

·   ·   ·

For fifty years, the banner of Public Affairs Press was carried by its owner Morris B. Schnapper, who published Gandhi, Nasser, Toynbee, Truman, and about 1,500 other authors. In 1983, Schnapper was described by *The Washington Post* as "a redoubtable gadfly." His legacy will endure in the books to come.

Peter Osnos, *Founder*